AF380001

World Nutritional Determinants

World Review of Nutrition and Dietetics

Vol. 45

Series Editor
Geoffrey H. Bourne, Grenada, West Indies

S. Karger · Basel · München · Paris · London · New York · Tokyo · Sydney

World Nutritional Determinants

Volume Editor
Geoffrey H. Bourne, St. Georges University School of Medicine,
Grenada, West Indies

11 figures and 34 tables, 1985

S. Karger · Basel · München · Paris · London · New York · Tokyo · Sydney

World Review of Nutrition and Dietetics

Vol. 42: Nutrients and Energy
G.H. Bourne, Grenada, West Indies (ed.)
XII + 228 p., 27 fig., 44 tab., hard cover, 1983. ISBN 3–8055–3710–7
Vol. 43: Nutrition, Food and Drug Interactions in Man
G. Debry, Nancy (ed.)
X + 202 p., 33 fig., 46 tab., hard cover, 1984. ISBN 3–8055–3800–6
Vol. 44: Nutritional Considerations in a Changing World
G.H. Bourne, Grenada, West Indies (ed.)
X + 218 p., 18 fig., 41 tab., hard cover, 1984. ISBN 3–8055–3837–5

National Library of Medicine, Cataloging in Publication
    World nutritional determinants/
    volume editor, Geoffrey H. Bourne.
    – – Basel; New York: Karger, 1985.– –
    (World review of nutrition and dietetics; v. 45)
    Includes bibliographies and index.
    1. Nutrition   I. Bourne, Geoffrey H. (Geoffrey Howard), 1909–   II. Series
    W1 W0898 v. 45 [QU 145 W9278]
    ISBN 3–8055–3948–7

Drug Dosage
    The authors and the publisher have exerted every effort to ensure that drug selection and dosage set forth in this text are in accord with current recommendations and practice at the time of publication. However, in view of ongoing research, changes in government regulations, and the constant flow of information relating to drug therapy and drug reactions, the reader is urged to check the package insert for each drug for any change in indications and dosage and for added warnings and precautions. This is particularly important when the recommended agent is a new and/or infrequently employed drug.

Advisory Board

Contents

Nutritionally Beneficial Cultural Practices

Christine S. Wilson, San Francisco, Calif. 68

Vitamins and Immunocompetence

Richard S. Panush, Gainesville, Fla.; *Jeffrey C. Delafuente*, St. Louis, Mo. . . . 97

Vitamin E and Blood

Nutritional and Hormonal Requirements of Mammalian Cells in Culture

Mechanism of Conversion of β-Carotene into Vitamin A –
Central Cleavage versus Random Cleavage

J. Ganguly, P.S. Sastry, Bangalore 198

Wld Rev. Nutr. Diet., vol. 45, pp. 1–41 (Karger, Basel 1985)

Optimum Nutrition through Better Planning of World Agriculture

Fred A. Kummerow[1]

Burnsides Research Laboratory, University of Illinois, Urbana, Ill., USA

Contents

In this paper I will discuss four parameters related to improving world-wide nutrition: (1) the role of calories and nutrients; (2) the role of food choices in the development of degenerative diseases; (3) the limits of agriculture in providing calories and essential nutrients, and (4) the role of government in better planning of world agriculture.

[1] I would like to thank Ms. *Donna Stowe* for her secretarial and editorial assistance.

I. The Role of Calories and Nutrients

Conception of the Need for Calories

Although human milk is the first source of calories to which almost everyone is exposed, other sources are soon utilized in the diet of the individual. In the developed countries, a wide array of food items, from cereals to canned and frozen foods, are readily available to provide calories in optimum amounts. In many of the developing countries the mother is forced to use whatever is available to provide supplementary calories and hopefully enough essential nutrients to sustain life. The lack of success in the developing countries is shown by comparison of vital statistics among countries (table I). Health professionals in developing countries have lowered these infant mortality rates. However, it has been estimated that one quarter of the human population goes to bed hungry every night [*Crittenden,* 1982]. In a world in which all countries are becoming increasingly aware of the basic needs of their citizens, far too few governments assign sufficient resources for providing an adequate diet or for making available means for population control. The latter is highly important and basic to a solution. My discussion in this article, however, shall be limited to nutrition and food production. What can be done within the framework of present day knowledge of nutrition?

Health professionals, it seems to me, have failed to educate the population in both the developed and the developing countries to the need for calories to sustain life. Our enzyme systems operate at 98 °F, well above room temperature, and therefore need heat from an outside source. The major sources of heat are provided by calories from fats, sugars, and cereals. These calories are essential to sustaining life. For individuals in developed countries to avoid calories by spending billions of dollars on noncaloric drinks and low calorie food items, often of no nutritional significance, is a poor use of funds and resources. At present, 70% of the $ 3 billion per year's worth of saccharin produced in the world is going into soft drinks [*O'Sullivan,* 1983]. Because saccharin is carcinogenic and its use is banned in some developed countries, substitutes are being developed. For example, it is anticipated that the sales of aspartame, a sweetener prepared from α-phenylalanine, will increase from $ 14 million in 1981 to $ 500 million by 1986. The current price for α-phenylalanine is $ 27/lb. It would seem more economical to use sugar at 25c/lb and realize that the calories supplied by sugar do not need to be supplanted by more expensive calories from low-nutrition snack foods.

Table I. Chances per 1,000 live births of death from infective and parasitic diseases in 1974–1978 [WHO Health Statistics Annual, 1980]

	♂	♀		♂	♀
Developed countries			*Developing countries*		
Australia	5.6	5.2	Costa Rica	35.4	31.8
Canada	5.9	5.2	El Salvador	109.8	92.9
Denmark	4.9	3.9	Mauritius	43.0	41.0
United Kingdom	4.8	3.7	Singapore	56.7	26.4
USA	7.8	7.3	Cuba	20.5	14.6

If Americans and Europeans had a better understanding of the role of calories in the diet, they might manage them with less weight gain and have more money in their pockets for something more worthwhile than low calorie food and drink. Calories are feared because when we consume more than the number needed to keep our enzyme systems operating at 98 °F, the excess calories are converted to body fat and stored for future need. An excess of approximately 110 cal/day (less than one 12-oz bottle of soft drink) adds 12 lb of weight per year. Most people prefer to keep the slim look of youth and not add more weight. They therefore believe that drinking a noncaloric drink provides an answer to the weight problem. It would certainly be more healthful to drink milk or fruit juices, which ounce for ounce are less expensive and more nutritious, and reduce the fat and sugar content in the total diet by 110 cal/day [*Kummerow,* 1979a].

People of the developing countries also have a problem with calories but it is usually directly opposite to that of the developed countries: it is often a shortage of calories. Some do produce surplus calories in the form of sugars, vegetable oils and other agricultural products. These are often exported for foreign exchange to import cereal grains (table II), to pay off interest on international loans, or to buy armaments; but that is another story. The lack of purchasing power in large segments of the population in the developing countries does not allow for the purchase of sufficient cereal grains to provide the calories needed. Much of the malnutrition in developing countries, particularly marasmus, a wasting disease, has been blamed on a lack of protein, but calories are also crucial. As *McLaren* [1974] has pointed out: 'Marasmus results from grossly restricted intake of all nutrients and energy. Dietary factors, especially in marasmus, are of second-line order of importance, and in a multifactorial aetiology, poverty, ignorance,

Table II. Imports and exports (in 100 metric tons) for developing and developed countries in 1980

	Import	Export
Cereals		
Developing		
Central America	15,994	1,187
Developed		
USA	1,925	1,129,058
Sugar		
Developing		
Central America	8,286	81,605
South America	8,906	39,162
Developed		
USA	37,208	141
Canada	8,749	6,401
Coconut and Palm Oil		
Developing		
Oceania (Fiji, Papua-New Guinea)	–	801
Asia (Malaysia, Indonesia, Philippines)	4	36,206
Developed		
North America	6,584	187
Western Europe	12,237	1,602

Data from: 1980 Yearbook of International Trade Statistics [1981]; 1980 Food and Agriculture Organization Trade Yearbook [1981]; Food and Agriculture Organization Monthly Statistics Bulletin [1982].

bad housing, poor hygiene, and lack of family planning all conspire. Food-consumption data and dietary surveys incriminate [an] energy [deficit] rather than [a] protein deficit. Increasing the energy intake and not ... [the] protein [intake] has produced catch-up growth in undernourished children. Lack of nutriment ... with an energy gap rather than a protein gap is the crux of the matter; but how to match the intake of the child with its requirements remains a problem of puzzling complexity.'

Conception of the Need for Nutrients

McLaren [1974] in his article entitled 'The Great Protein Fiasco' stated, 'We still do not know, for instance, whether or not many of us are actually eating levels of protein that are harmfully high. The experts now talk of the "safe level of protein intake" but have been unable to set its upper limit. The recommended dietary protein intake has been progressively lowered by the experts in recent years so that it is now about where it was nearly 70 years ago.' The Food and Nutrition Committee of the National Research Council (NRC) presently recommends 56 g of protein per day and has defined the optimum daily intake of vitamins and minerals for humans (table III).

The daily intake of protein, vitamins and minerals has also been defined by the NRC for pets, poultry, cattle and swine. It is relatively simple to follow these recommendations for animals with diets of cereals and legumes. Such diets have resulted in healthier animals and decreased market maturation time. For example, swine fed a diet of ground corn and soybean meal supplemented with minerals and synthetic vitamins reach maturity in 6 months instead of 2.5 years. A bred gilt will give birth to a piglet weighing 2–4 lb. Within 6 months this piglet fed the supplemented corn soybean diet weighs 220 lb.

Millions of human beings in the developing countries use ground corn and beans as their major source of calories and nutrients. They are shorter of stature than Americans and have a shorter life expectancy (table IV). A ton of commercial swine ration contains 1,725 lb of ground corn, 220 lb of soybean meal and 55 lb of a lysine vitamin mineral mix. This ration compares favorably with a diet meeting the NRC recommendations for man (table III). However, without the supplemental 55 lb of minerals and vitamin mix per 2,000 lb, the commercial ration does not provide the same rate of growth. Is it possible that the diet of millions of people eating a diet composed largely of ground corn and beans could be improved by simply adding a mineral vitamin supplement to the ground corn? Since 1944 in the southern states of the USA, ground corn (corn grits) has been supplemented with niacin, riboflavin and thiamine to prevent pellagra. Such a simple supplementation decreased the death rate from pellagra in the USA from 2,000 in 1941 to 12 in 1944 [Vital Statistics of the United States, 1941, 1944].

Interestingly, the commercial mineral vitamin mix used in swine rations differs somewhat in trace mineral composition from the mineral mix recommended by the NRC (table III). The commercial mix is designed for swine kept in confinement under crowded conditions which foster stress, competition for feed, and a tendency to 'eat like pigs'. The difference that a nutritionally complete diet might have on the behavior of people living under crowded conditions has yet to be studied. Certainly the nutri-

Table III. Recommended daily allowances (RDA) for man and swine compared

	RDA for man[1]	RDA for swine[2]	Actual commercial swine ration pre-mix[3]
Protein, g	56	67.5	
Minerals			
Calcium, mg	1,200	2,300	50,000
Phosphorus, mg	1,200	1,800	22,500
Sodium, mg	*[4]	250	37,500[5]
Chlorine, mg	*	330	
Potassium, mg	*	750	
Magnesium, mg	400	100	350
Iron, mg	18	38	450
Zinc, mg	15	25	45
Manganese, mg	*	1	
Copper, mg	*	1.5	
Iodine, mg	150	0.04	10
Selenium, mg	*	0.04	
Vitamins			
Vitamin A, µg β-carotene	6,000	2,200	27,272
Vitamin D, IU	400	55	6,818
Vitamin E, mg	10	2.8	18.75
Vitamin C, mg	60	*	
Vitamin K, mg	*	50	
Thiamin, mg	1.4	0.33	
Riboflavin, mg	1.7	0.75	4.55
Niacin, mg	18	5.5	34.09
Vitamin B_6, mg	2.0	0.38	
Folacin, µg	400	150	
Vitamin B_{12}, µg	3.0	5.5	0.036
Pantothenic acid, mg	*	3.3	11.36
Choline, mg	*	275	227.27
Biotin, mg	*	0.03	

[1] American males 15–18 years old weighing 66 kg.
[2] Swine weighing 1–5 kg live weight.
[3] Pre-mix added to the commercial corn-soybean ration.
[4] Not given.
[5] This number represents the total amount of sodium and potassium chloride.
Data from: National Academy of Sciences [1980]; Nutrient Requirements of Swine [1979].

Table IV. Life expectancies in developed versus developing countries: percent expectation of life at age 75

Developed countries		Developing countries	
North America		*South America*	
United States	10.4	Uruguay	9.1
Canada	10.3	Chili	8.8
Scandinavia		*Central America*	
Sweden	9.7	Bermuda	8
Norway	9.6		
		Asia	
Central Europe		Singapore	7.4
The Netherlands	9.7		
Switzerland	9.7	*Africa*	
		Mauritius	7
Asia			
Japan	9.5	*Mediterranean*	
Israel	9.3	Malta	6.6
Mediterranean			
Greece	9.5		
Spain	8.9		
Italy	8.6		
Yugoslavia	8.4		

tional well-being in over-populated countries of people on corn and bean diets could be improved by adding the proper mineral vitamin mix to the corn meal. Enriched food would be far more economical than weapons for the USA to supply.

The cereal based diets in some developing countries are also supplemented with beans as a protein source. In India, rice and wheat are supplemented with a large variety of beans, sometimes called pulses, and in China these cereals are supplemented with soybeans. In India and China, a small amount of animal protein is used to supplement the vegetable protein. In India, buffalo milk and in China, pork represent the main animal protein sources. The milk is so important in India that cows are considered sacred and are not slaughtered for meat. The Chinese consume only 13 g of protein and 142 cal as compared to 71 g of protein and 742 cal/capita from meat per day for Americans [*Eberstadt,* 1979]. The Chinese also consume only 76 cal of visible fat (margarine, shortening and frying oils) as compared to 576 cal of fat/capita/day for Americans or Northern Europeans. Americans and Europeans also consume more calories from sugar than Chinese do, approximately 4 times the calories [FAO Trade and Production Yearbooks, 1980].

The diets of people in the developing countries therefore differ from those in the developed countries. People in developing countries have a minimum of 'empty calorie' sources in their diet and require fewer nutrient rich foods to attain the daily NRC recommended levels of protein, minerals and vitamins. People in the developed countries need to eat the protein, vitamin and mineral rich eggs, meat and dairy products in order to compensate for the 'empty calories' from fat and sugar. If a corn-soybean diet that contains no extra calories from sugar and fat still needs to be supplemented with vitamins and minerals for optimum health and growth in swine, it seems obvious that a diet which contains 1,884 cal from sugar and fat out of a total of 3,000 cal/day (such as the American diet) must be supplemented with protein, vitamins and minerals or must contain foods that provide these nutrients.

Some health professionals in the developed countries have recommended diets that contain less meat, eggs and dairy products [*Blackburn*, 1980]. They do not seem to have considered the limitations that their recommendations would put on food choices. Substitute foods equal in nutritive value to meat, eggs and dairy products are not readily available in the required quantities in the supermarkets of the developed countries. Furthermore, sugar and fats are an economical source of calories and are used as essential components of highly popular foods such as snack foods, french fries, cookies, cakes and doughnuts. An increased consumption of 'empty calories' requires an increased intake of protein, vitamins and minerals from milk, meat, eggs, fruit and vegetables.

Although it is possible to construct a diet free of meat that would meet NRC requirements by using bread, vegetables, fruits, skim milk and eggs [*Levine and Parker,* 1979], the amount of vegetables, fruits and milk that would be required for everyone to become a vegetarian is not available [*Kummerow,* 1979b] for 218 million Americans 365 days/year. A sample diet of vegetables, fruits, nuts, eggs and three 8-oz glasses of skim milk/day would require 69.5 gallons of milk per capita/year or 6.7 times more than the 10.2 gallons per capita/year actually being produced. Even if all the whole milk now available for the production of cheese and other dairy products would become available as fluid milk there would still be a shortage of 39.9 gallons per capita. To supply three 8-oz glasses of skim milk per capita daily, production would have to be increased by 2.4 times which is only possible if the price of milk could increase enough to bring more farmers into milk production. The same is true for the production of peanuts and walnuts. The use of 10 g of walnuts and 40 g of peanuts/day as suggested by *Levine and Parker* [1979] adds up to 18.25 kg per capita or 9 times more than the total US production of nuts including peanuts. The climate is too severe in the USA Midwest to replace corn and soybeans with either peanuts or walnut trees. Furthermore, neither corn nor soybeans could be ground and sold as such because of the rancidity that would quickly develop and thus result in their rejection by most Americans.

The greater consumption of fruits and vegetables is desirable because they are excellent sources of vitamins, minerals, and fiber. They are usually abundantly available in supermarkets in the quantities that we now use them in the USA. However, we do not even approach the production that would be required for our total population to be largely vegetarian. I seriously question the agricultural and economic feasibility of increasing the production of 4.5 times the amount of all fresh fruits and vegetables now produced in the USA. We could increase production but the supply of vegetables is limited by considerations not encountered with livestock and poultry. Because of soil and climate, vegetables for human use can be grown economically only on a relatively small share of the available farmland. Their planting and harvesting requires a great deal of seasonal hand labor. After harvesting, which must be accomplished very quickly and within a very restricted time span, these perishable foods must be used or processed (by canning or freezing) within a few days. However, animal production, because less seasonal, has the important advantage of flexibility. The corn and soybeans grown for animal feed on millions of acres of midwest farmland are easily stored in huge grain elevators without processing and are easily used only as needed. Moreover, there is considerable leeway in deciding when animals should be sent to market: beef or pork need never flood the market at any one time. Due to such practical considerations, only the livestock, dairy, and poultry industries can supply the population's nutritional needs with the efficiency, economy, and abundance we now enjoy. Implementing dietary changes which would decrease the intake of animal protein to one half of present amounts would require a radical redistribution of crops, a vastly expanded supply of seasonal farm labor, and an enormous increase in food processing capacity.

Perhaps the diet recommended by the Food and Nutrition Board of the National Research Council is not as yet, fully 'balanced'. The recommended level of vitamins, minerals, and protein (the RDA) may be present if all the

Table V. Comparison of results in swine fed the basal ration or the basal ration plus 5 or 10% used fat for 3 months [from *Kummerow* et al., 1974, unpublished data]

	Basal	Basal + 5% fat	Basal + 10% fat
Live weight (average)	214	261	247
Dressed weight (average)	159	200	187
Lost weight (average)	55	60	60
Standard deviation	19.9	14.6	26.6
Weight gain/animal	105	152	138
Feed consumption/animal	225	346	256
Feed efficiency ratio	0.42	0.43	0.54

Swine averaged 109 lb at 3 months; slaughtered at 6 months.

food items in the diet carry their full complement of RDAs. When white bread, however, is substituted for whole wheat bread, french fried potatoes for baked potatoes, cooked weiners for beef roast, a soft drink for milk, and cakes or pies rather than fruit as a dessert, RDAs are not likely to be present in the desired amounts. The recommendations as listed do not fully consider the negative influences of over-consumption of sugar and fat. The RDA of 56 g of protein from a diet composed of whole cereal, baked potato, fresh vegetables and fruits, milk, meat, and eggs may be sufficient. The typical American diet, however, of french fries, soft drinks, and high caloric desserts may require even more protein than the 104 g/day/capita presently consumed. Eliminating nutrient-deficient foods would automatically increase the nutrient value of the diet. It is possible to 'balance out' the American diet by increasing the consumption of nutrient-packed foods and decreasing intakes of the less nutritious foods.

The significance of the level of fat in a well-balanced diet, as measured by weight gain, is brought out with animal models such as swine. The addition of 5 lb of fat to a balanced diet of ground corn and soybean meal containing more than the RDA for weanling swine increased their weight approximately 50 lb more at 6 months of age than those fed this diet without additional fat. The addition, however, of 10 lb of fat to the basal diet (without increasing the protein, vitamin and mineral level) caused approximately 15 lb less weight gain than those fed 5 lb of fat (table V). There is, therefore, an optimum ratio of calories to protein and other essential nutrients in animal diets which may be paralleled in human diets. Although 56 g of protein/day is presently considered sufficient by the National Research Council, the present consumption of 104 g of protein/capita/day in the USA may represent a better calorie-protein ratio and sufficient vitamins and minerals in foods as they are actually consumed. Even though adults may be able to live for a number of years on vegetarian diets constructed from food items that are available in a supermarket, these food items would have to contain a low percentage of sugar and fat for the total diet to meet NRC requirements. Furthermore, cereals and vegetables have a low protein energy density, that is, protein calories per serving. For young, rapidly growing children, low protein energy density makes it difficult for them to eat enough protein: combining egg or milk with vegetable proteins is one way to cancel out deficiencies in vegetable protein. The growth lag from poor nutrition does not need to be drastic. Studies in Bangladesh showed that 1- to 4-year-old children with a mid-arm circumference 75% that of average healthy children in developed countries had a death rate 4.5 times higher than children in developed countries [*Cravioto* et al., 1966].

The nutrition protocol which is presently applied by clinical physicians to severely ill and to geriatric patients has shown that a normal level of serum albumin is possible on a proper mixture of vegetable proteins, but that such a vegetable mixture of protein does not supply all of the trace elements [*Kaminski,* 1982]. These trace elements are not present in sufficient amounts in corn or vegetables. The daily intake of cereals and beans may furnish an adequate protein and calorie intake for humans on low fat, low sugar vegetable protein diets, but such diets may be marginal in vitamins and the trace minerals. If one decreased consumption of animal protein from 70 to 35 g/day so as to decrease cholesterol intake from 600 to 300 mg/day and the calories of fat from 1,104 to 720 (42–30% from fat) without substituting plant protein of equivalent nutritional value to animal protein, the calories per gram of protein could increase from 15 to 20.5. What such food choices have on the calorie-protein ratio and the development of degenerative diseases has not been considered by the well-meaning advocates such as the American Heart Association (AHA) and the National Institutes of Health (NIH) who recommend a decrease of cholesterol intake from 600 to less than 300 mg/day by curtailing the consumption of meat, eggs and dairy products to one half of present consumption levels.

These advocates assume that the approximately 1,500 mg of cholesterol synthesized from carbohydrate or fat each day, primarily in the liver, is sufficient to provide the body's needs for cholesterol. As cholesterol is synthesized from acetyl-CoA, the primary building block obtained from sugar, starches, oils, fats or even protein through a complex system of enzymatic reactions, it is assumed by these advocates that dietary sources of cholesterol are unnecessary. In fact, they believe that dietary sources of cholesterol add to the in vivo cholesterol in the serum and that cholesterol in the serum adds to cholesterol deposits in the arterial wall. Some segments of the food industries in the developed countries have taken advantage of such an assumption by advertising their products with slogans such as 'with no cholesterol'. I consider such slogans a disservice to the food industry: they cause confusion in understanding the very complex metabolic process that causes cholesterol deposits in the arterial wall, a process known as atherosclerosis which is commonly held to be responsible for 90% of all coronary heart disease (CHD).

II. The Role of Food Choices in the Development of Degenerative Diseases

The majority of studies on the role of food choices in the development of degenerative diseases such as atherosclerosis have focused on the transport of lipids in the blood stream and the possible deposition of these lipids into the intimal layer of the coronary arteries [*McGill,* 1979]. According to this commonly held idea, the LDL lipoproteins infiltrate the endothelium layer of cells, the first layer of cells next to the lumen, and invade the intima. In response, smooth muscle cells in the media layer migrate to the intima and phagocytize the LDL lipoproteins. The residual lipid is believed to gradually form fatty streaks which are found in the arteries of all humans and do not seem to interfere with arterial function. The next stage in the development of atherosclerosis does interfere with blood flow through the arteries: the continual migration of the smooth muscle cell into the intima is believed to gradually thicken it to the point of 'plaque' formation. The plaques (largely composed of cholesterol) are believed to be 'hardened' by the infiltration of calcium and become a threat to the integrity of the arteries. In the small coronary arteries that furnish the heart muscle with blood, the plaques can diminish blood flow to the point that the blood supply to the heart muscle is compromised enough to cause CHD. In the larger arteries such as the aorta, the rhythmic pulsations of the artery are believed to cause tiny tears between the uninvolved artery and the plaque which allows blood to leak out of the arteries (an aneurysm).

My own 40 years of research deal with the role of lipids in nutrition. The subject is complex and involves more than the study of dietary cholesterol and serum lipoprotein levels. We need to understand what is happening at the cell membrane level. I disagree with the AHA and NIH recommendations regarding lowering the dietary cholesterol from 600 to 300 mg/100 ml/day.

Healthy People, the Surgeon General's Report on Health Promotion and Disease Prevention [1979] stated: 'Prevention is an idea whose time has come.' The report also stated that: 'Premature heart disease is unequivocally associated with elevated blood cholesterol levels. Stroke risk, too, is increased by elevated serum cholesterol although the association is not as strong as for heart disease ... direct evidence from animal studies supports the linkage of atherosclerosis with high levels of fats (particularly saturated) and cholesterol in the diet.'

On the other hand, Toward Healthful Diets [1980], a report from the Food and Nutrition Board of the NRC National Academy of Sciences, Washington, D.C., stated: 'No significant correlation between cholesterol intake and serum cholesterol concentration has been shown in free-living persons in this country.' How can the official pronouncements of two governmental agencies, the Surgeon General's Office of the Department of Health, Education and Welfare on the one hand and the Food and Nutrition Board of the National Academy of Sciences on the other, present such opposite points of view on the leading cause of death in the USA? Both agencies used experts to help them reach their decisions. How is it possible for experts to study the same data and yet come to such opposite conclusions?

This can happen easily, for the available data lends itself to opposing conclusions. On the one hand, experts that hold the surgeon general's point of view assume that dietary sources of cholesterol elevate blood cholesterol levels sufficiently to enhance the rate at which the LDL lipoproteins diffuse through the endothelium and accumulate within the arterial wall. On the other hand, experts that hold to the Food and Nutrition Board point of view believe that the amount of cholesterol in the diet is not high enough to elevate the blood cholesterol level significantly and is not responsible for the accumulation of cholesterol within the arterial wall.

Two sources of data have been used in the attempt to implicate dietary sources of cholesterol such as meat, eggs and dairy products in the development of atherosclerosis (1) epidemiological or clinical studies, in which the diets of various population groups were compared with the incidence of heart disease, and (2) the data obtained when animal models are fed cholesterol. Neither the epidemiological and clinical studies nor the animal model data has successfully implicated cholesterol at present dietary levels as the cause of atherosclerosis.

Epidemiological and Clinical Data

Epidemiological studies have involved comparisons of the diets in developed and developing countries with the incidence of CHD. These studies were interpreted to indicate that population groups such as Northern Europeans on diets which contain animal protein as furnished by meat, eggs and dairy products have a higher rate of death from CHD than Southern Europeans or people from the developing countries on diets which contain largely beans and cereals as a protein source. I do not agree with this interpretation, although WHO statistics indicate that Finns, Swedes, and Germans have a higher death rate from CHD than Italians, Greeks, Spaniards, Indians, etc. [WHO Health Statistics Annual, 1980]. However, the death rates from other causes than CHD have not been considered. When they are, one comes to a different viewpoint than the one held by the Surgeon General.

In countries such as Italy and Greece, the death rate from cancer is higher than in Finland and Germany, and the death rate from cerebral strokes is higher in Japan than in the USA or Germany. However, the percentage of people alive at age 74 is approximately the same in the USA as in Italy. Furthermore, as population groups in developing countries have a lower life expectancy than those in developed countries, they do not live long enough to reach the age at which degenerative diseases develop. It may be correct to state that more people in Northern European countries die of CHD than those in Southern European countries. However, one should add that an avoidance of animal protein does not increase longevity.

Furthermore, it is simple for a physician to assign the cause of death to CHD in the absence of an obvious cancer or life-threatening metabolic disease. A heart that is no longer pumping blood will result in death regardless of the underlying cause for its malfunction. According to *Geller* [1983], only 9.9% of the more than 950,000 deaths due to major cardiovascular disease and 11.2% of the more than 385,000 deaths due to malignancy were autopsied in the USA in 1977. Moreover, the accuracy of certification of the underlying cause of death in 2,557 cases autopsied in various hospitals indicated that the underlying cause of death had been inaccurately recorded in 42% of the cases. As medical care varies from country to country, data based on causes of death without an autopsy could be in error.

Nor can we even say in such epidemiological studies that cholesterol consumption was the major variable. In order to lend credence to the observation that population groups on high animal protein diets suffer from more CHD than those on low animal protein diets, various clinical studies have been carried out [National Diet-Heart Study Research Group, 1968; *Levy* et al., 1979]. To date, none of them has been able to prove that decreasing the intake of animal protein (cholesterol and saturated fat) will decrease the incidence of CHD. For example, the most recent and best financed study to date, the Multiple Risk Factor Intervention Trial (MRFIT), was undertaken for an average of 7 years at 22 Medical Centers to ascertain whether modification of elevated serum cholesterol levels by dietary changes, treatment of hypertension and lowering cigarette smoking would result in reduction of death from CHD [Multiple Risk Factor Intervention Trial Research Group, 1982]. Eligibility for the study was determined at three successive screenings of 361,662 men, 35–57 years of age, resulting in 12,866 men judged free of heart disease, diabetes mellitus, and diastolic blood pressure and serum cholesterol levels lower than 115 mg Hg and higher than 350 mg/dl. They were randomly assigned to either a special intervention (SI) program consisting of stepped-care treatment for hypertension, counseling for cigarette smoking and dietary advice for lowering blood cholesterol levels, or to a usual care (UC) group, receiving the usual sources of health care in the community. The first contracts were awarded by the NIH in 1972 and screening was completed 28 months later. Eating patterns for the SI group recommended a decrease in saturated fat to less than 10% of calories and a cholesterol intake of less than 300 mg/day which was lowered to 250 mg/day in 1976. The UC group was given no particular dietary advice.

As of February 28, 1982, there were 260 deaths among the UC men and 265 among the SI men of which 145 and 138 were ascribed to cardiovascular causes, respectively. The death rate from all causes was therefore 2.1% higher in the SI men which was deemed statistically insignificant. The number of deaths was substantially short of expectation and was interpreted to be caused by: (1) the recent reduction in CHD mortality in the USA, the reasons for which are not understood; (2) the stringent exclusion screening resulting in a selection of men with lower than expected mortality; (3) the phenomena of lower than expected mortality in clinical trials involving human volunteers, and (4) the substantial risk factor changes in the UC as well as the SI men.

Table VI. Clinical data concerning patients used in the present study [from *Taura* et al., 1977]

Non-atherosclerotic intima					Atherosclerotic intima				
case No.	age	sex	choles-terol mg/ 100 ml	blood pressure	case No.	age	sex	choles-terol mg/ 100 ml	blood pressure
1	43	M	234	104/62	3	59	M	229	124/60
2	48	M	247	130/90	9	42	M	180	128/88
7	54	M	230	115/70	14	49	M	311	134/72
8	53	M	180	128/88	15	56	M	222	145/90
10	63	M	215	140/80	17	66	M	300	110/70
11	59	M	294	130/90	22	60	M	275	140/85
12	45	M	171	115/75	23	60	M	220	130/85
16	40	M	312	140/85	24	57	M	272	140/90
18	61	M	229	150/90	25	31	M	272	160/110
19	48	M	277	150/80	26	34	M	219	130/80
20	50	M	206	148/78	27	60	M	339	140/90
21	60	M	248	110/70	29	52	M	248	130/85
28	60	M	295	110/70	33	45	M	265	120/70
30	69	M	285	150/85	35	45	M	282	145/95
31	41	M	246	126/90	36	65	F	250	130/90
32	65	M	199	172/84	38	60	M	245	120/70
34	58	M	221	130/80	39	47	M	213	110/65
37	59	F	241	120/70	40	62	M	208	126/80
42	59	M	208	160/100	46	56	M	190	126/70
43	38	F	275	120/85	48	55	M	200	135/85
44	62	M	200	130/85	49	52	M	167	148/80
45	67	M	217	145/95	51	52	M	254	130/90
47	70	M	266	170/100					
50	54	M	242	130/90					
52	54	M	270	180/108					
53	59	M	232	130/90					
54	47	M	271	150/105					
55	59	M	195	130/80					

In the absence of dietary advice, one could assume that the diet of the UC men contained the customary level of saturated fat and 600 mg cholesterol/day, although actual dietary cholesterol levels were never measured by the MRFIT researchers. Because animal food products such as meat, milk and eggs serve as the major source of cholesterol, their consumption at customary levels by the UC men brings doubt to the idea that these sources of nutrients serve as dietary risk factors. In my opinion, the additional meat, milk and eggs necessary to supply the 350 mg of cholesterol over and above the recommended 250 mg of cholesterol/day for the SI men would have added substantially to the available nutrients in the diet of UC men. It is unfortunate for human nutrition that so much attention has been focused on dietary sources of cholesterol as potential risk factors, while their contribution to total nutrition and their potential contribution to general good health and longevity have been ignored.

Animal Models

Although the Surgeon General may be justified in stating that *premature* heart disease is associated with elevated serum cholesterol levels, the majority of men and women with proven CHD did not have serum cholesterol levels over 260 mg/100 ml (table VI). Of 50 proven cases of CHD subjected to coronary by-pass operations only 17 had serum cholesterol levels over 260 mg/100 ml [*Taura* et al., 1977]. High levels of saturated fat and cholesterol in the diet do enhance the development of atherosclerosis in an animal model but such an experimental protocol does not reflect the actual dietary condition of Americans or Northern Europeans whose diets contain a sufficient level of unsaturated fats to cancel out the serum cholesterol-elevating tendency of saturated fats [*Kokatnur* et al., 1958; *Kummerow* et al., 1960]. In fact, due to an increase in the amount of linoleic acid (18:2ω6) in margarine and shortenings a shift in the polyunsaturated-saturated fatty acid ratio has occurred and may have contributed to a general lowering of the serum cholesterol level.

The fat processing industry has been aware of the need for an increase in the linoleic acid content of margarines and shortenings. Since 1965, the percentage of linoleic acid in Crisco shortening, for example, has increased from 8 to 24% and the percentage in margarines from 8 to 27% (table VII). Because of an increase in the consumption of poultry, the consumption of linoleic acid from animal fats has increased from 2.8 to 3.5 lb/capita/year since 1909. On a daily basis, this amounts to a total of 164 g of fat/capita of which 30 g is linoleic acid or 19% of calories which is more than ample to meet NRC requirements [*Yamanaka* et al., 1980].

The amount of cholesterol in the diet at levels presently furnished by meat, milk and eggs is also not relevant to the serum cholesterol level. *McGill* [1979], in an appraisal of cholesterol as a causative factor in atherogenesis, stated that the 600 mg/day of cholesterol in the average American

Table VII. Composition of margarines and shortenings[1] [courtesy of Proctor & Gamble, 1968]

	Shortenings		Margarines	
	new	old	new[2]	old[3]
Percent saturated fatty acids	25	27	19	21
Percent monounsaturated fatty acids	47	61	50	63
Percent polyunsaturated fatty acids	28	12	31	17
Percent *cis-cis* linoleic	24	8	27	8
Percent *trans* acids	20	30	27	44

[1] % of total fatty acids.
[2] Higher unsaturated margarines.
[3] Lower unsaturated margarines.

diet of 2,400 cal contributed about 14 mg/dl to the total serum cholesterol level and that if the intake were reduced to 300 mg/day the reduction in serum cholesterol would be about 7 mg/dl, a difference which is not significant to the development of pathology. *Jackson* et al. [1977] found that a dietary cholesterol level (0.4%) equivalent to 40 eggs/day for 16 weeks did not alter the structure and composition of swine lipoproteins, although *Mahley* et al. [1978] have reported the appearance of a very low-density lipoprotein and high-density lipoprotein C fraction in serum of swine fed a dietary cholesterol level (1.5%) equivalent to approximately 150 eggs/day. No Sudan IV stainable fatty streaks were noted in the aorta of swine fed 0.4% cholesterol; however, fatty streaks and mitosis have been noted at higher dietary cholesterol levels. A dietary level of 37,500 mg of cholesterol (equivalent to 150 eggs/day) does contribute to the development of atherosclerosis in swine, but such an amount is not relevant to the 600 mg/day of cholesterol in the diet of North Americans or Northern Europeans.

Some researchers believe, however, that 600 mg/day of dietary cholesterol may influence atherogenesis even though it does not alter the serum cholesterol concentration significantly. In order to understand this point of view, which I do not hold, it is necessary to consider how water-insoluble compounds such as cholesterol, fat (triglycerides) and phospholipids are 'carried' in the blood. The reasons for my disagreement with this view can then be shown by surveying studies on amino acid imbalance, ApoE synthesis and serum hormone levels.

Table VIII. Composition (%) of lipoproteins [from *Kummerow,* 1979a]

	Total		Lipid[1]			
	protein	lipid	triglyceride	phospholipid	cholesterol	cholesterol ester
VLDL	10	90	50	20	8.5	8.5
LDL	25	75	10	30	10	50
HDL	50	50	10	50	5	25
VHDL[2]	60	40	12	76	1	8

[1] Unesterified fatty acid.
[2] Very high density lipoproteins.

The 'carrier' is made up of a complex series of amino acids of 'apoproteins' which form a water soluble micelle with the cholesterol, triglycerides and phospholipids. The apoproteins are synthesized in the intestinal tract and the liver from both the essential and nonessential amino acids. As the intestinal tract synthesizes only a limited amount of apoprotein, however, the micelle of lipid-carrying protein (lipoprotein) that is formed in the intestinal wall is very limited in protein content. These micelles can be separated by centrifuging into various fractions of different fat content (table VIII). The major (chylomicron) fraction contains very little protein. However, as it is circulated through the liver, the chylomicron is stripped of its fat and cholesterol by lipoprotein lipase, and the lipoproteins of increasing protein content are secreted into the blood by the liver.

The fatty acid composition and possibly the spatial configuration of the micelle in the serum are influenced by the mixed fatty acid composition of the dietary fats. In the presence of an excess amount of polyunsaturated fatty acids, less cholesterol is needed to keep the micelle of serum lipoproteins in a fluid state. In the presence of saturated or isomeric fatty acids, more cholesterol is needed to keep the micelles of lipoproteins in the fluid state. For example, the two margarines used in the National Diet-Heart Study [1968] differed in isomeric as well as linoleic acid content, and their consumption resulted in different serum cholesterol levels.

The lipoprotein can be fractionated by ultracentrifugation or electrophoresis into five fractions: chylomicrons, very low density lipoproteins

(VLDL), low density lipoproteins (LDL), high density lipoproteins (HDL), and very high density lipoproteins. The percentage of lipid components gradually decreases from 90% in the VLDL to approximately 50% in the HDL fraction or class. Each lipoprotein class is heterogeneous with respect to its apoprotein constituents. ApoA refers to the apoproteins that are primarily, but not exclusively, found in HDL. ApoB is the major apoprotein of LDL, but also comprises about 35% of the protein in VLDL. ApoC represents a group of proteins originally described in VLDL, but which are also present in HDL [*Jackson* et al., 1976].

The apoproteins that have been sequenced (characterized) to date contain varying numbers of the essential amino acids interspersed among the amino acids that can be synthesized in vivo. Their arrangement may be important to their lipid-carrying capacity [*Jackson* et al., 1974]. Approximately 50% of the amino aicds in the apoproteins cannot be synthesized in vivo and must, therefore, be furnished in the daily diet. An imbalance in dietary amino acid level can increase serum cholesterol levels [*Kokatnur and Kummerow*, 1961]. For example, an imbalance caused by 4% additional lysine increased serum cholesterol levels in chickens to higher levels than 0.5% dietary cholesterol (equivalent to 50 eggs/day) or 302 and 193 mg/100 ml, respectively [*Schmeisser*, 1983]. These studies indicate that dietary sources of cholesterol are not required to increase the serum cholesterol level.

Amino acid imbalance may also be responsible for the variable effects of animal and vegetable proteins on cholesterol metabolism that have been shown in rabbits, rats, swine and other mammals [*Kritchevsky* et al., 1982]. It has recently been shown in both rabbits and rats that casein feeding increased cholesterol absorption and decreased steroid excretion compared with the soy diet; on the other hand, rats fed the amino acid mixtures simulating casein and soy showed no differences in steroid excretion, but casein amino acids increased cholesterol synthesis compared to soy amino acids [*Huff and Carroll*, 1980]. It has been suggested that the arginine-lysine ratio of these amino acids may be responsible for the effects of these proteins and amino acid mixtures on cholesterol metabolism [*Kritchevsky* et al., 1982].

The high serum cholesterol and high triglyceride levels that have been reported in 1 out of 250 heterozygous and 1 out of 1,000 homozygous humans [*Levy* et al., 1979] may be the result of an imbalance in serum estrogen, androgen and progesterone levels. Previous workers [*Stamler*, 1971] have treated patients with signs of CHD with estrogen on the basis that women do not develop CHD as readily as men. However, without first assaying the plasma for its estrogen level, the plasma triglyceride and cholesterol level could be further increased rather than decreased after the administration of estrogen. Furthermore, the increase may be greater in someone with already high lipid levels as compared with someone of low lipid levels. This was found to be the case in the RO nonlayers as compared with normal 4-month-old pullets when both were injected with estrogen (table IX). The lipid levels of the RO nonlayers increased much more dramatically than in the 4-month-old pullets [*Hagan* et al., 1983].

At 6 months of age the plasma levels of triglycerides and cholesterol in the nonlayer (RO) hens ranged from 2 to 3 times higher than the comparable layers [*Toda* et al., 1980]. Serial sectioning revealed atherosclerosis in both left and right coronary arteries from 6-month-old nonlayers but not from the layers [*Tokuyasu* et al., 1980]. When 0.5% estrogen was fed for 8 weeks to 5-day-old New Hampshire-Columbian cross female chicks, their plasma cholesterol levels were approximately 900 mg/100 ml higher than those on the basal diet and 600 mg/100 ml higher than those fed 1% cholesterol [*Leszczynski* et al., 1982]. The increase in plasma cholesterol level due to estrogen was partially counteracted with testosterone and progesterone.

'Errors' in the assembly of the apoproteins may also be responsible for the accumulation of cholesterol in the liver. For example, in the apoE isolated from the serum of humans suffering from type III hyperlipidemia cysteine is substituted for arginine at position 158 [*Rall* et al., 1982]. The apolipoprotein E (apoE) may be responsible for the defective binding of apoE. It has been postulated that apoE is important in the lack of uptake of chylomicron remnants in the liver which leads to higher serum lipid levels [*Sherrill* et al., 1980]. Studies concerning the method of cholesterol and triglyceride transport in the blood stream are crucial to the views of *McGill* and others who believe that serum cholesterol is involved in the development of atherosclerosis. However, such studies do not fully explain the mechanism through which cholesterol is deposited in the arteries.

For example, rabbits fed 2% cholesterol and lanthanum trichloride did not develop lesions to the same extent as those fed 2% cholesterol alone, despite their serum cholesterol levels being similar [*Kramsch* et al., 1980, 1981]. Similar results were noted in rabbits fed 2% cholesterol and the calcium 'blocker' nifedipine [*Henry and Bentley,* 1981]. Therefore, the presence of an excessive amount of cholesterol in the serum is not as crucial to the development of lesions as the selective ability of the cell membrane to

Table IX. Mean plasma hormone and lipid values from hormone-treated female chickens [from *Hagan* et al., 1983]

Treatment group[1]	n	Days of treatment	E_2 pg/ml	P ng/ml	TG mg/dl	PL mg/dl	C mg/dl
20-month-old layers							
Control	5	0	107 ± 75	2.72 ± 1.69	$1,248 \pm 481$	939 ± 218	111 ± 29
		7	176 ± 27	2.03 ± 1.76	$1,166 \pm 601$	835 ± 283	119 ± 36
		14	346 ± 161	1.07 ± 0.76	$1,228 \pm 758$	727 ± 414	188 ± 72
Estradiol 1 mg/kg b.w.	6	0	190 ± 60[2]	1.76 ± 1.17	$1,303 \pm 891$[3]	847 ± 345[3]	109 ± 41[3]
		7	720 ± 70*	1.79 ± 0.69	$4,550 \pm 2,587$*	$2,369 \pm 1,153$*	292 ± 95*
		14	$1,055 \pm 457$*	3.27 ± 2.67	$9,381 \pm 2,007$*	$4,282 \pm 830$*	786 ± 263*
Progesterone 4 mg/kg b.w.	5	0	220 ± 106	1.51 ± 1.50	$1,626 \pm 694$[1]	$1,000 \pm 271$[3]	136 ± 31
		7	90 ± 44*	4.41 ± 3.59	$1,273 \pm 565$	589 ± 155	251 ± 99
		14	227 ± 102	3.76 ± 2.74	205 ± 51*	369 ± 42	172 ± 25
4-month-old pullets							
Control	6	0	163 ± 80	0.185 ± 0.081	176 ± 45	214 ± 26	105 ± 11
		7	228 ± 81	0.922 ± 0.770	170 ± 41	197 ± 17	109 ± 7
		14	254 ± 39	1.25 ± 0.53	248 ± 51	250 ± 26	130 ± 10
Estradiol 1 mg/kg b.w.	7	0	361 ± 247[2]	0.185 ± 0.103	188 ± 62[3]	235 ± 43[3]	106 ± 10[3]
		7	$3,365 \pm 4,081$	0.777 ± 0.702	761 ± 261*	515 ± 147*	203 ± 48*
		14	$1,292 \pm 615$*	2.62 ± 1.94	$1,282 \pm 562$[2]	869 ± 309*	266 ± 81*
Progesterone 4 mg/kg b.w.	7	0	241 ± 108	0.344 ± 0.333[3]	163 ± 43	229 ± 31	109 ± 14
		7	$1,279 \pm 1,654$	1.69 ± 0.98	189 ± 27	259 ± 12*	135 ± 13*
		14	305 ± 102	3.42 ± 1.57*	230 ± 51	302 ± 35*	142 ± 14

E = estrogen

P = progesterone

TG = triglyceride

PL = phospholipid

C = cholesterol

b.w. = body weight

[1] Treatments were given once each day; controls were injected with propylene glycol. All possible paired t-tests within groups and 2-tailed t-tests between hormone treatment and control groups with the same days of treatment were calculated.

[2] Within this group, 2 of 3 possible paired t-tests were significantly different (p < 0.05).

[3] Within this group, all 3 possible combinations of paired t-tests were significantly different (p < 0.05).

* Hormone-treated group value is significantly different (p < 0.05, 2-tailed t-test) from comparable control group value.

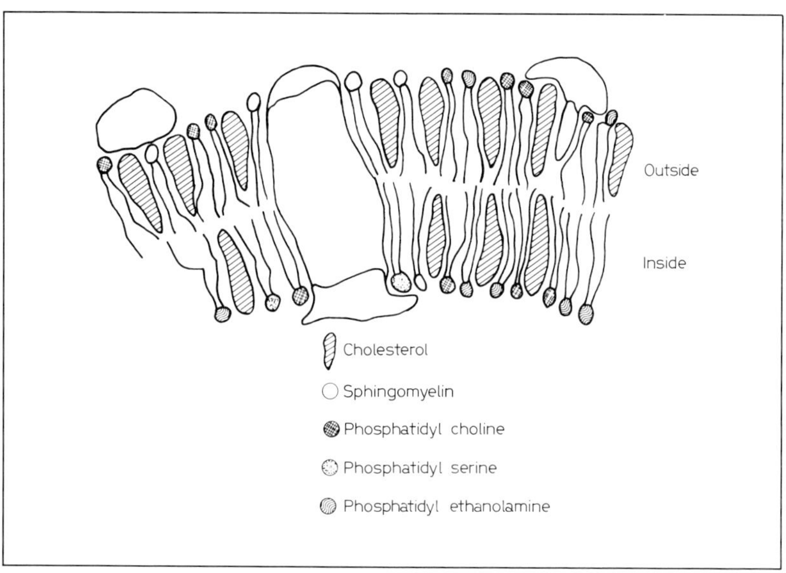

Fig. 1. Erythrocyte membrane.

allow the flow of cholesterol and calcium into the intima cells. However, as lanthanum or nifedipine are not diet components, the question remains, how can the flow of cholesterol and calcium into the intima be controlled and atherosclerosis prevented? To understand these complex relationships it is necessary to consider this question at the molecular level.

Modification of Cell Membrane Composition by
Dietary Lipids and Its Implications for Atherosclerosis

Everyone of the trillions of cells that make up the body share the same structural components. They are encapsulated in a membrane composed primarily of protein and lipid (phospholipid and cholesterol) and second-arily of carbohydrate (fig. 1). Membranes are essential cellular components which envelop the cell and divide it into various functional compartments. They maintain a permeability barrier between these compartments, trans-mit information, and allow the selective movement of nutrients across them. The carbohydrate groups are attached to lipids or proteins and are localized mainly at the cell surface. They play a vital structural role in determining receptor and antigenic sites. Proteins may be either peripher-ally associated with membranes or may be an integral component whereby proteins usually span the membrane. They function as recognition or catalytic sites. Cholesterol is the predominant sterol in animal cell membranes and its main functional role identified to date is the modification of the motional freedom of fatty acyl chains [*Demel and DeKruyff,* 1976]. When added to membranes consisting of lipids with gel-like molecular arrangements, it increases their motional freedom. Conversely, it reduces the mobility of acyl chains in a fluid state [*Singer and Nicholson,* 1972]. A change in any of these essential structural components may precipitate a disease process [*Wallach,* 1973]. Increasing significance now is being attached to the role of membrane lipids, both directly in influencing membrane properties and indirectly through their effect on membrane proteins. The phospholipids in the membranes, through the hydrophobic character of the fatty acids attached to them, constitute the predominant structural element of membranes and impart to them an impermeability to hydrophilic molecules. To form a membrane the fatty acid hydrocarbon chain must be of a length within certain limits, generally containing between 12 and 24 carbon atoms. Biological membranes in general contain acids with 16, 18 or 20 carbon atoms. Fatty acid unsaturation is another key factor related to membrane function. Its principal effect appears to be to create fluid environments in the membrane, so that proteins and lipids can diffuse in the plane of the membrane and protein can undergo a conformational change during interaction with ligand, substrate or effector.

A side view of the Singer and Nicholson model (fig. 2) indicates that the integral protein that spans the membrane is wedged between fatty acid chains and serves as a conduit for potassium, sodium, calcium, and other nutrients into the cell [*Klausner* et al., 1980]. The oleic, linoleic, and arachidonic acid fatty acid chains at the β-position in the phospholipids of animal tissue all have a *cis* configuration which provides a 'kink' in the chain. As each double bond adds one 'kink' to the spatial configuration, the linoleic and arachidonic acid can 'curl up' and provide more space between the fatty acid chains than would be the case if all of the fatty acids were in a straight chain. One could speculate that the curled-up polyunsaturated fatty acids act as a spring to keep the protein in place and yet act like an accordion in order to accommodate the spatial configuration to as close a 'normal' situation as possible. Even the substitution of a three (eicosatrienoic acid) for a four double bond (arachidonic) fatty acid may influence spatial configuration in phosphatidylcholine [*Walker and Kummerow,* 1964].

Two new sources of substrates, both influential in the integrity of cell membranes, were introduced into the American diet in 1920: hydrogenated

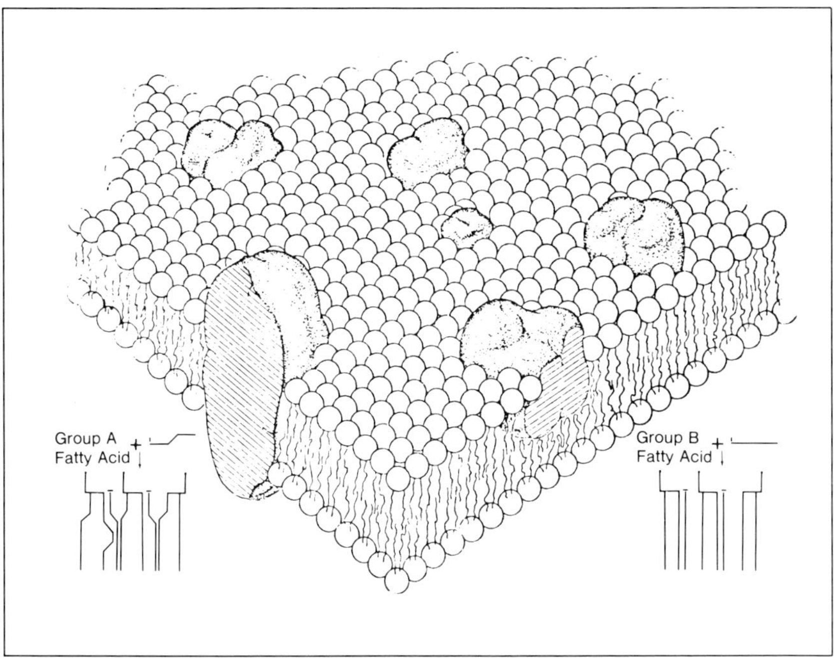

Fig. 2. Singer/Nicolson model of cell membrane [*Singer and Nicolson,* 1972]. *Karnovsky* [1979] classified the PUFA into 'A' and saturated and *trans* fatty acids into 'B'. He believed that the group A fatty acids (FFAs) partition into fluid domains and group B FFAs partition into gel-like domains and that the group A FFAs cause a shift in membrane-bound calcium from protein calcium-binding sites to the lipid phase.

fats and dietary sources of vitamin D. Hydrogenated fats became available as shortenings and margarine in the USA in 1920 [*Swern and Bailey,* 1964] as a replacement for lard, tallow and butter. They are produced by catalytic hydrogenation of vegetable or fish oils and in the process some of the double bonds in the unsaturated fatty acids are 'isomerized', i.e. converted from a *cis* to a *trans* geometric form or the double bond shifts along the carbon chain. Isomerization changes the spatial configuration so that the *trans* form of oleic acid (elaidic acid) displaces a greater area than stearic acid but less than oleic acid (fig. 3). Therefore, the substitution of elaidic acid for palmitic or stearic acid at the α-position would increase and the substitution of elaidic for linoleic acid at the β-position would decrease the spatial configuration of a phospholipid in the membrane.

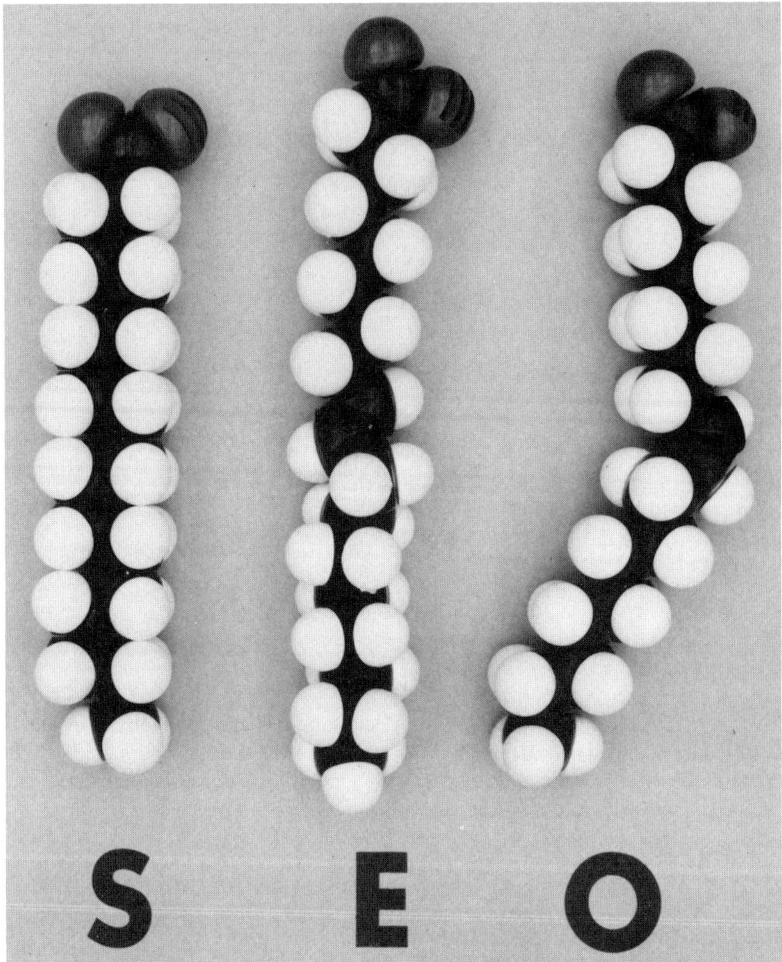

Fig. 3. Spatial configuration of S-Stearic, E-Elaidic *(trans),* and O-Oleic *(cis)* acids.

Recently, *Fontaine* et al. [1981] demonstrated that the uptake of radioactive calcium was enhanced in the epithelial cells isolated from the intestinal wall of chicks fed vaccinic acid (the *cis* 11 monoenoic acid) as compared to the epithelial cells isolated from the intestinal wall of chicks fed *trans* vaccinic acid (the *trans* 11 monoenoic acid). *Rasmussen* et al. [1982] believed that a greater percentage of *cis* unsaturated fatty acids in the phosphatidylcholine in the membrane of the intestinal wall columnar epi-

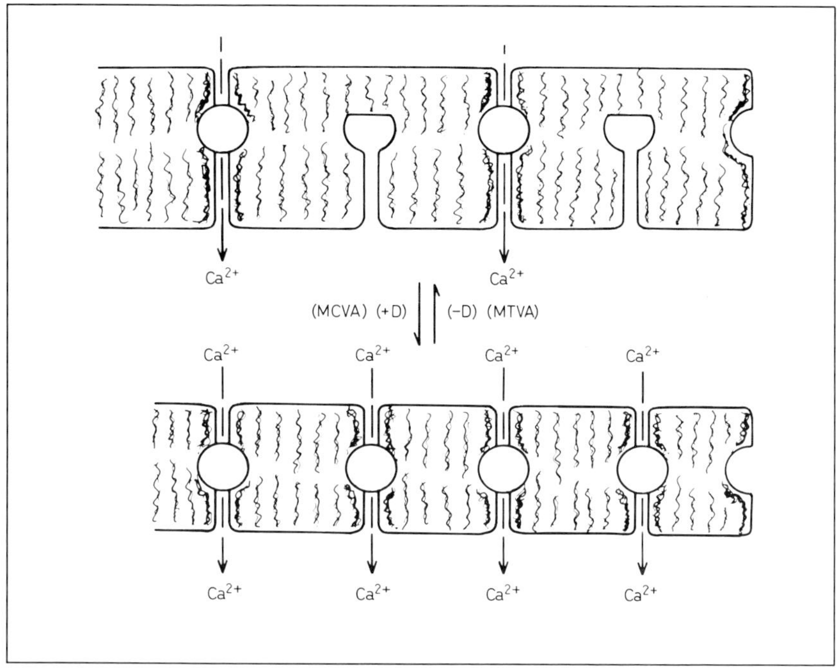

Fig. 4. A possible model of action suggested by *Rasmussen* et al. [1982] by which the methyl esters of *cis-* and *trans*-vaccenic acid (C 18:1, 11Δ) (MCVA and MTVA) added in vitro produce changes in calcium transport across the luminal membrane of the enterocyte similar to those seen after administration of 1,25-(OH)$_2$-D$_3$ in vivo. In the proposed model, either manipulation (MCVA in vitro or 1,25-(OH)$_2$-D$_3$ in vivo) leads to an increase in membrane fluidity, which in turn converts cryptic calcium channels in the membrane into active ones. Conversely, addition of MTVA in vitro reverses the effect of 1,25-(OH)$_2$-D$_3$ by decreasing membrane fluidity. Printed by permission of *Rasmussen* et al. [1982].

thelial cells would increase membrane fluidity which in turn might convert the cryptic calcium channels to active channels and thus increase calcium uptake by the cells (fig. 4). In the presence of an excess amount of polyunsaturated dietary fat, *Rasmussen* et al. [1982] believed that a larger percentage of linoleic acid is incorporated into the phospholipids in the cell membranes. In swine fed a diet supplemented with hydrogenated fats, a higher count of degenerated cells was noted in the coronary arteries than those on the fat-unsupplemented diet [*Kamio* et al., 1979]. However, in swine fed a diet supplemented with hydrogenated fat and an adequate amount of linoleic acid, less degenerated cells were noted in the coronary arteries [*Royce* et al., 1983]. Furthermore, the coronary arteries of swine fed a diet supplemented with lard, an animal fat, contained approximately the same amount of degenerated cells as those fed a high oleic safflower or hydrogenated soybean oil, both vegetable fats. The synthesis of prostaglandins also requires a sufficient amount of linoleic acid to provide an adequate balance of eicosanoids [*Kinsella* et al., 1981].

The studies of *Rasmussen* et al. [1982] involved epithelial cells of the intestinal tract and not the intimal cells in the arteries. If one applies this observation to the complex series of events that result in atherosclerosis, one may be able to explain why saturated fats, cholesterol and vitamin D all accelerate the process. The saturated fats may cause the cell membrane to be less fluid and may stimulate the incorporation of more cholesterol into the membrane causing it to be too fluid and allowing too much calcium to flow into the intimal cell. That vitamin D has a similar affect was shown in swine fed saturated fat with and without vitamin D [*Kamio* et al., 1977]. However, where sufficient linoleic acid, protein, vitamin, minerals and a minimum amount of vitamin D was present in the diet, less atherosclerosis was present in the arteries [*Royce* et al., 1983]. An animal fat such as lard was no more atherogenic than a vegetable fat such as safflower oil or hydrogenated soybean oil when included in a diet with all required nutrients present in adequate amounts.

A sufficient amount of dietary linoleic acid must also be present to provide a balance of platelet-aggregating thromboxane and fluidizing prostacyclin in the serum [*Kinsella* et al., 1981]. In the absence of linoleic acid more thromboxane is present in the serum [*Sullivan and Mathias,* 1982] which could result in more thrombus formation in the blood. The increase in the linoleic acid content of the dietary fats since the early 1960s may have resulted in less thrombosis in the developed countries during the 1970s and 1980s.

There is no question that essential fatty acids are required for the synthesis of the eicosanoids such as the thromboxanes, prostaglandin and leukotrienes. Their ratio of synthesis seems to be influenced by the presence of *trans* fatty acids. Better control of hydrogenation has decreased the presence of such *trans* acids in margarine and shortenings which may have some influence on the rate of thrombus formation in the coronary arteries and the rate of CHD. An adequate amount of linoleic acid seems to 'balance out' the level of cholesterol in the serum as well as control the fluidity of the cell membranes.

*The Role of Dietary Factors Other than Cholesterol and
Saturated Fat in the Development of Atherosclerosis*

Through their effect on membranes, the various oxidized sterols that are present in foods may be more significant than cholesterol in influencing the initiation and development of atherosclerosis. 25-Hydroxycholesterol and 7-ketocholesterol are auto-oxidation products of cholesterol and may arise during processing or cooking of cholesterol-containing foods. The potential for these derivatives to exist in the food chain and to produce pathological effects may be ascertained by performing some simple calculations. If the average American consumes 500 mg of cholesterol per day, and if 0.1 % of that were oxidized, a very small proportion, this would result in the consumption of 500 µg of oxidized derivatives of cholesterol. If 20 % of that were 25-hydroxycholesterol, it would correspond to the ingestion of 100 µg 25-hydroxycholesterol per day. I believe that the toxicity of vitamin D is related to the formation of 25-hydroxyvitamin D [*Holmes and Kummerow*, 1983], the main circulating form of vitamin D, which shares several common structural features with 25-hydroxycholesterol. Toxic responses to vitamin D have been reported with doses as low as 50 µg per day which stresses the relevance of cholesterol oxidation [*Kummerow*, 1983].

Another oxidized sterol, 7-ketocholesterol, is quite potent in increasing the frequency of degenerated smooth muscle cells of the aortas of experimental chicks (table X). Other workers have confirmed that 25-hydroxycholesterol is similarly angiotoxic [*Breslow* et al., 1975]. In the former study it was not determined whether significant Ca^{2+} deposition occurred in the coronary arteries of those fed 7-ketocholesterol because the techniques employed would have permitted leaching of Ca^{2+} from deposition sites. The question of maintaining Ca^{2+} deposition in thin sections in a way that reflects in situ deposition is one being addressed in several laboratories but with no clear resolution as yet. The hypothesis that one of the principal toxic effects of these sterols is caused by their insertion in cellular membranes needs to be pursued. This insertion may influence the permeability properties of membranes, particularly the permeability of plasma membranes to Ca^{2+} because a large electrochemical gradient for Ca^{2+} exists across these membranes.

Similar principles may apply during hypervitaminosis D where 25-hydroxyvitamin D is the main circulating metabolite. It too may influence the permeability properties of membranes following its insertion in the membrane. Arterial calcification and smooth muscle cell necrosis are typical features of hypervitaminosis D [*Taura* et al., 1979].

Table X. Smooth muscle cell degeneration in abdominal aorta after either 4 or 8 weeks dietary treatment [from *Toda* et al., 1982]

Dietary Group	Type of degeneration		Combined frequency		Total number of cells counted
	cyto-lysis	pyk-nosis	total count	frequency %	
4-week treatment [1]					
Basal	18	9	27	0.7	3,830
1% cholesterol	21	4	61	1.9	3,149
7-Ketocholesterol[2]	42	283	325	9.8	3,315
1% cholesterol + 7-ketocholesterol[2]	108	357	465	13.2	3,524
8-week treatment [3]					
Basal	31	13	44	1.2	3,659
1% cholesterol	48	32	80	2.1	3,786
7-Ketocholesterol[2]	224	491	715	22.0	3,248
1% cholesterol + 7-ketocholesterol[2]	269	465	734	23.6	3,110

[1] χ^2 test for significant difference in group mean combined frequencies ($p < 0.05$) of degeneration: 1% Chol. + 7-KC > 7-KC > 1% Chol. > basal.
[2] 7-Ketocholesterol was force-fed once daily, other dietary components were fed ad libitum.
[3] χ^2 test of combined frequencies of degeneration ($p < 0.05$): 1% Chol. + 7-KC, and 7-KC > 1% Chol. > basal.

Oxidized sterols may therefore be much more important to the modification and phagocytosis of the smooth muscle cell than LDL lipoproteins. The modified smooth muscle cell in the intima is known to be more susceptible to phagocytosis than the smooth muscle cell in the media. The flow of calcium into the smooth muscle cell that is accelerated by the insertion of oxidized sterol into the cell membrane may become toxic to the cell and cause its conversion to a degenerated cell. These cells may gradually accumulate as debris that contains cholesterol, cholesteryl esters, phospholipids, and calcium.

In summary, excessive amounts of dietary cholesterol, the amount and balance of saturated and unsaturated fat, the presence of a 'balanced' amount and kind of protein and amino acids, the presence of an optimum amount of vitamins and minerals, the presence of oxidized sterols and the balance of sex hormones all influence serum lipid levels. Furthermore, in the presence of optimum amounts of protein, vitamins and minerals, the addition of either animal or vegetable fat did not increase the atherogenic influence of the diet as judged by intimal thickening and cell degeneration.

III. The Limits of Agriculture in Providing Calories and Essential Nutrients

We will now leave the problem of what goes wrong in the human body and turn to the world-wide problem of the lack of adequate food for human nourishment. Agriculture is defined as the cultivation of land as in the raising of crops, husbandry, tillage farming, etc. Unfortunately, only 15% of the world's land can be cultivated for crops [*Crabbe and Lawson, 1981*]. Of the rest, 5% is devoted to horticulture, 2% to urban development, 20% to forests, 10% to deserts and 45% to stock raising. The rapid expansion in urban development is almost all at the expense of cultivated land. Further expansion of land for cultivation is not likely in the foreseeable future. It therefore seems necessary to use stock raising areas to maximize efficiency in the production of nutrients in the developing countries.

In the USA, the efficiency of stock raising areas has been improved by taking advantage of the protein residue in the by-products of two major industries: the residue left after the extraction of cottonseed oil from cottonseed and the residue left after the extraction of oil from soybeans. These residues are pressed into small cubes and are fed to cattle or ground and mixed into chicken or swine rations. These residues are also shipped to Northern Europe for use in animal rations; however, they are not in surplus supply and are therefore not available to other stock raising areas of the world.

In the USA, the calves that are born on the open range in Kansas and Colorado in March are transferred to weather-protected feed lots or wheat pastures in late fall and are fed hay or allowed to graze on winter wheat until they reach 675 lb in weight. Whole corn and defatted soybean meal are used to supplement cattle in order to add another 200 lb of meat to the protein that is already present in the 675-lb feeder cattle fed grass. Feeder cattle are grown to 675 lb in weight on land that is not fit for wheat, corn or soybeans but at 675 lb are not at their most economical weight; that is, more of the animal is available as meat if an additional 200 lb of muscle is added to it by feeding it corn, soybean meal and essential vitamin and mineral supplements in feed lots in addition to grass and hay. Grasslands in the developed countries are either fertilized to provide increased forage yields or are interlaced with alfalfa or clover to increase the protein content of the forage. Also, in current USA feed lot operations, the energy content of the rations is increased by the addition of molasses and fat and a decrease in the percentage of hay in order to stimulate a rapid weight gain. Under such conditions beef cattle are brought to market at 18 months of age. The cattle on grasslands in the developing countries are presently not being supplemented with grain or protein or sugar-fat supplements. Thus, in the developing countries a newly born calf takes 5 years to reach market weight in comparison to only 1.5 years for the supplemented cattle in the USA.

A typical animal in a feed lot gains 3 lb/day by consuming 16 lb of hay, 1 lb of soybean meal and 4 lb of ground corn each day. Thus, in order to add 200 lb of weight, 268 lb of corn and 65 lb of soybean meal is consumed. The extra 200 lb of meat added in the feed lot furnish more of the essential vitamins, trace minerals and usable protein than the 268 lb of corn and the 67 lb of soybean meal would furnish if they were consumed directly by man as primary protein sources.

A farmer in Illinois can produce more protein and calories per acre by growing corn or soybeans and at a higher profit per acre than by growing a cereal crop such as wheat. The average yield of wheat, corn and soybean/acre in Illinois was 38, 111 and 33 bushels (in 1978) respectively, and provided a net profit of $21, $69 and $98/acre, respectively. This 1 acre of wheat, corn and soybeans would provide 262, 548, and 644 lb of protein and 3.3, 10.1 and 3.6 million calories, respectively, if consumed directly. If consumed by meat-producing animals or poultry (as is the case for 85% of the corn and 90% of the soybean crop), a loss of approximately 25% of the protein would occur. However, the acre of corn or soybeans would still provide more food as meat, milk and eggs than wheat or: 191, 466 and 485 lb of protein and 2.5, 7.9 and 2.7 million calories, respectively [*Watt and Merrill, 1975*].

To recommend the direct consumption of wheat, corn and soybeans ignores at least two factors. One, some cereals like hybrid corn yield more 'nutrition' when consumed by animals which yield high quality animal food products, such as beef, than when eaten as a cereal or flour. The present up-grading of our grassland with corn and soybean crops is primarily responsible for 'balancing out' the increase of 'empty' fat and sugar calories in the American diet. If wheat farming took over in the Midwest, as some prominent nutritionists have advocated, we could become as deficient in calories and protein as the major developing countries in which cereals furnish the bulk of protein and calories.

Secondly, the yield of protein and calories that can be obtained from an acre of land depends on its fertility and rainfall. Climatic and soil conditions in much of Northern Europe limits food production to dairy farming: the rainfall and poor soil limits farming to the growth of grass. Other areas of the world, such as the tropics, are limited to sugar cane and palm or coconut trees. Such trees produce 3 times more oil per acre than soybeans. Thus, by taking maximum advantage of climate and soil conditions, most areas of the world can produce either excess proteins (dairy products) or excess calories (sugar and vegetable oil). If the products from this agriculture were shared more equally, everyone would benefit. The surplus protein in the form of milk powder and cheese in the developed countries could be shipped to the developing countries with the hope that a more equal distribution of calories and nutrients would result.

Agriculture is also limited by available labor. The unindustrialized developing countries have labor available for the planting and harvesting of vegetables and so subsist on a largely vegetarian diet. The developed countries have less farm labor and so must use an animal-based agriculture.

Although the 'state of the art' in food technology has contributed substantially to nutrition in the developed countries, it has not had as great an impact in the developing countries. The identification of the vitamins as essential nutritional factors and their synthesis has made it possible to add them to the rations of swine and poultry and thus greatly increase the supply of meat and eggs in the developed countries. This technology could be modified so as to supply the missing nutrients in the ration of livestock in the developing countries. For example, in addition to ground corn, synthetic amino acids or urea, vitamin A and trace minerals, could be made available to these underfed livestock. Although chickens are available in most countries, their productivity could be greatly increased by a choice of genetic stock and by trace nutrient supplements to maximize egg production. In each case, 'tailor-made' supplements for whatever limiting dietary factors could be added to maximize egg production or rapid growth.

IV. The Role of Government in Better Planning of World Agriculture

For agriculture to meet the nutritional needs of the world population better planning is required. What is needed is planning that takes into consideration both the needs of the land and of the population. In order for agriculture to remain productive for future generations as well as for the present inhabitants of the earth, a world agriculture association must come into being that would have more economic power than any organization presently in existence. Such an organization would require the best computer technology available in order to catalog week by week the total amount of grains, meats, oils, fats, sugar, dairy products, fruits and vegetables produced that week and make them available as needed by the world population. Production must be determined by ecology, i.e. there should be no attempt to plow up grassland in order to try to grow wheat, nor should rain forests be cut down to grow a few more acres of soybeans. Wet lands and swamps should not be drained merely to add a few acres of land to produce a crop already in surplus. The present practice of raping the land for the short-gain of an individual should be considered a crime.

Each community can catalog its assets and then work out methods for which the particular piece of land is best suited. Desert areas in tropical zones could be made liveable by bringing in water from water surplus areas. The push to expand housing on cultivated land should cease; less valuable land could be used for housing. At present, many housing developments are carried out by real estate developers with no thought in mind except profit and with no regard for the best use of the land. Real estate developments are also very provincial in scope, i.e., each country thinks only of the real estate under its jurisdiction. The tropical islands in the Carribbean have benefitted from real estate development, but this development must also take the native population into consideration. Thousands of Mexicans swam across the Rio Grande River in the belief that the USA would provide them a better life. Yet the possibility that hundreds of thousands of acres of land in Northern Mexico could be provided with water and made productive for agriculture needs consideration. Northern Africa could provide a productive living area for those seeking mild climates and for the Arabs who are so desperate for a homeland. Modern technology can make this possible.

The Israelis have shown how deserts can be made productive. Their technology could make most of North Africa a liveable area. It would take electric power to pump the necessary water, but this power could be made available through solar energy. New means to desalt the water of the Mediterranean could be used. If oil can be pumped for thousands of miles of pipelines, so can water. In fact, as water is so important to agriculture, its uses should be under the jurisdiction of both agriculture and urban administrators [*Talbot,* 1977, 1978].

Agriculture must also develop much closer ties with food processors in every country. In the USA, there is a working relationship between the livestock industry and the manufacturers of the feed supplements such as vitamins and amino acids. As a result, poultry and eggs are readily available at economical prices. Both canning and freezing fruits and vegetables take more heat and power than is available in some areas of the world. During World War II, the US government encouraged the use of small community canning units. Such units could be developed in many countries presently in short supply of vegetables in winter. Presently, few cans or glass jars of any kind of fruit or vegetable are available in the large cities of Eastern Europe. Growing vegetable gardens could be encouraged.

A new approach is needed to the surplus grain problems in the USA as well as to the problem of inadequate nutrition for hundreds of millions of people presently unable to purchase grain because of lack of funds. Better planning of world agriculture would have made unnecessary the present attempt to increase the price of corn to a level above its production cost in the USA. The present plan is to decrease the production of corn from 8.397 billion bushels in 1982 to 5.640 billion bushels in 1983. It is hoped that a difference of at least 1.8 billion bushels of corn between 1982 and 1983 will increase the price of the 1983 crop sufficiently to provide a profit to the farmer. As 85% of the corn is fed to livestock any increase in the price of corn will result in an increase in the price of meat. The consumer will still pay for the increase in costs although not in the form of taxes for federal subsidies to food growers. Donating the corn or bartering it for rare metals with the developing countries would have provided a better means of reducing the corn surplus: because it could have been fed to cattle, it would not have been in competition with grains used for human consumption.

I suggest that the WHO or some other organization with the necessary funds purchase corn for areas of the world willing to cooperate in feeding studies which would develop methods for bringing the cattle to market weight sooner. Thus we could improve grassland efficiency in these areas. The meat from these cattle should then be made available to the people in these areas so as to improve their nutritional status. Currently, in Central America, cattle are often fattened on lots owned by large US corporations. In Central America, for example, 80% of their meat production is shipped to the USA. The amount of meat in the diet of people in the developing countries could be increased by not exporting this beef for foreign exchange. If diets in these countries were improved by increased meat protein intake, the need for the costly health programs funded by the developed countries might be reduced with the result that overall costs would not increase. It seems unnecessary to import 1 billion pounds of beef/year into the USA

Table XI. World need for metals in 1967–1969 and in 2000 AD [from *Lahiri,* 1976]

Country	Per capita consumption, kg			
	crude steel	aluminum	copper	zinc
Per capita consumption of metals in selected countries[1]				
Developing				
Brazil	51.0	0.90	0.512	0.482
Chile	71.0	–	2.140	–
India	12.0	0.23	0.083	0.151
Mexico	73.5	0.53	0.969	0.312
Developed				
Sweden	603.0	8.10	10.924	4.388
UK	404.5	6.83	9.646	4.959
USA	60.5	17.25	8.950	5.900
FRG	523.5	8.84	9.761	5.883

Region	Per capita consumption, kg			
	steel	aluminum	copper	zinc
Per capita consumption of metals by regions projected for the year 2000 AD				
Developing				
Africa	20	0.24	0.16	0.07
India: Low growth	26	0.51	0.20	0.32
High growth	51	0.98	0.44	0.70
Latin America	100	1.72	0.91	0.95
China	60	0.79	0.63	0.54
Developed				
Western Europe	710	20.24	10.50	6.83
Japan	1,450	45.86	21.40	13.19
USSR	850	17.47	7.84	3.92
USA	890	52.25	14.63	9.41

[1] Average values for the years 1967–1969.

from developing countries that are so desperately in need of nutrients for their own people [*Simpson,* 1982].

As previously mentioned, the developed countries with surplus grains might barter these grains for metals that are being produced in the developing countries. It has been estimated that the need for metals in 2000 AD will be greater than it was in 1982 (table XI). The consumption of aluminum, copper and zinc in the USA in 1969 has been predicted to increase from 17,

9 and 6 to 52, 15 and 9 lb/capita by 2000 AD. Many of the developing countries do have metals to export. It would be advantageous to have metals in adequate supply for every country.

The knowledge and technology are available to improve world agriculture and nutrition. We lack the social and political will to do so. Telecommunication and jet transportation brings the USA and Europe closer to China and India in travel time than San Francisco to New York in pre-World War II travel time. Television brings the world into every living room in the developed countries, yet has not broken down the barriers and fears of the average person for 'foreigners'. We must realize that no one in the world is a foreigner. The only foreigner would have to come from outer space.

Therefore, we must all change our thinking in regard to the people of other countries. Familiarity does not breed contempt, it is absolutely essential to survival. To survive we must experiment with economic systems until they no longer develop deficits.

So much of the world's resources are used in armaments that there is a scarcity of funds for promoting better nutrition. The ball is in the court of the politicians rather than in the court of the nutritionists. The ever-increasing sophistication of nuclear armaments has compressed the time in which the political leaders of the world can decide to end life or to create a more stable world. Those tens of millions who will be instantly burned into a vapor, such as the tens of thousands at Nagasaki and Hiroshima, will be more fortunate than those hundreds of millions that will die slowly of radiation sickness and starvation. In the museum at Hiroshima, everyone should see the coins suspended in the sheets of glass. As the people evaporated, the coins fell from their pockets into the melting concrete sidewalks. The over \$500 billion that are now spent on armaments each year could provide full employment and transform world agriculture so that everyone would have enough to eat. Our generation is destined to make a choice: the wrong one will end human life forever. We do have a choice, but we may be too full of distrust to make the right one.

Summary

The diets in both the developed and developing countries are dependent on agricultural practices, soil and climatic conditions. Diets made up largely of vegetable protein require that a substantial portion of the population be engaged in agriculture in order to plant and harvest the crops. The required labor is not available for the majority of the populations in the developed countries to exist on a vegetable protein diet. Furthermore, the mechanized agriculture in the developed countries has made it possible to develop a concentrated source of nutrients from animal sources so as to take advantage of calorie-rich economical crops, such as sugar cane and palm trees. This diet, when adequately prepared and 'balanced' so as to provide all of the various nutrients in adequate amounts, is not any more atherogenic or carcinogenic than a vegetable protein diet. A diet based on animal protein is not as monotonous to eat and is more palatable than a vegetable protein diet. Furthermore, because of its superior nutrient value, it provides for a greater resistance to infectious diseases and optimum metabolism of toxic factors in the liver. The suggestion that the saturated fat and cholesterol in animal food products are 'risk factors' in the development of coronary heart disease does not take into consideration the complex biochemistry that is basic to atherogenesis, the cause for 90% of coronary heart disease.

References

Blackburn, H.: Risk factors and cardiovascular disease; in The American Heart Association Heartbook (Dutton, New York 1980).

Breslow, J.L.; Lothrop, D.A.; Spaulding, D.R.; Kandutsch, A.A.: Cholesterol 7-ketocholesterol and 25-hydroxycholesterol uptake studies and effect on 3-hydroxy-3-methylglutaryl-coenzyme A reductase activity in human fibroblasts. Biochim. biophys. Acta *398:* 10–17 (1975).

Crabbe, D.; Lawson, S.: The World Food Book: An A–Z atlas and statistical source book (Kogan Page, London/Nichols, New York 1981).

Cravioto, J.; DeLicardie, E.R.; Birch, H.G.: Nutrition growth and neurointegrative development: an experimental and ecologic study. Pediatrics, Springfield *38:* 319–320 (1966).

Crittenden, A.: Hunger: the last word. It's poverty. RF Illustrated June (1982).

Demel, R.A.; DeKruyff, B.: The function of sterols in membranes. Biochim. biophys. Acta *457:* 109–132 (1976).

Eberstadt, N.: Has China failed? N.Y. Rev. *26:* 33 (1979).

Fontaine, O.; Matsumoto, T.; Goodman, D.B.P.; Rasmussen, H.: Liponomic control of Ca^{2+} transport: relationship to mechanism of action on 1,25-dihydroxy-vitamin D_2. Proc. natn. Acad. Sci. USA *78:* 1751–1754 (1981).

Food and Agriculture Organization Monthly Statistics Bulletin, vol. 5, pp. 1–75 (FAO, Roma 1982).

Food and Agriculture Organization Production Yearbook (WHO, Genève 1980).

Food and Agriculture Organization Trade Yearbook (WHO, Genève 1980).

Geller, S.A.: Autopsy. Sci. Am. *248:* 124–136 (1983).

Hagan, R.C.; Leszczynski, D.E.; Kummerow, F.A.: Comparative plasma lipid response of pullets and laying hens to estradiol and progesterone. Biochem. exp. Biol. (in press, 1983).

Healthy People: The Surgeon General's Report on Health Promotion and Disease Prevention. DHEW (PHS) Publication No. 79-55071 (US Dept. of Health, Educ. and Welfare, Public Health Service, Office of the Assistant Secretary for Health and Surgeon General, Washington 1979).

Henry, P.D.; Bentley, K.I.: Suppression of atherogenesis in cholesterol-fed rabbits treated with nifedipine. J. clin. Invest. *68:* 1366–1369 (1981).

Holmes, R.; Kummerow, F.A.: The relationship of adequate and excessive intake of vitamin D to health and disease. J. Am. Coll. Nutr. *2:* 173–199 (1983).

Huff, M.W.; Carroll, K.K.: Effects of dietary protein on turnover, oxidation and absorption of cholesterol, and on steroid excretion in rabbits. J. Lipid Res. *21:* 546–558 (1980).

Jackson, R.L.; Morrisett, J.D.; Gotto, A.M., Jr.: Lipoprotein structure and metabolism. Physiol. Rev. *56:* 259 (1976).

Jackson, R.L.; Morrisett, J.D.; Sparrow, J.T.; Segrest, J.P.; Pownall, H.J.; Smith, L.C.; Hoff, H.F.; Gotto, A.M., Jr.: The interaction of apolipoprotein-serine with phosphatidylcholine. J. biol. Chem. *249:* 5314 (1974).

Jackson, R.L.; Morrisett, J.D.; Pownall, H.J.; Gotto, A.M., Jr.; Kamio, A., Imai, H.; Tracy, R.; Kummerow, F.A.: Influence of dietary trans fatty acids on swine lipoprotein. J. Lipid Res. *18:* 182 (1977).

Kaminski, M.V., Jr.; Nasr, N.J.; Freed, B.A.; Sriram, K.: The efficacy of nutritional support in the elderly. J. Am. Coll. Nutr. *1:* 35 (1982).

Kamio, A., Kummerow, F.A.; Imai, H.: Degeneration of aortic smooth muscle cells in swine fed excess vitamin D_3. Archs Pathol. Lab. Med. *101:* 378–381 (1977).

Kinsella, J.B.; Bruchner, G.; Mai, J.; Shimp, J.: Metabolism of *trans* fatty acids with emphasis on the effect of *trans* octadecadienate on lipid composition essential fatty acids and prostaglandins: an overview. Am. J. clin. Nutr. *34:* 2307–2318 (1981).

Klausner, R.D.; Kleinfeld, A.M.; Hoover, R.L.; Karnovsky, M.J.: Lipid domains in membranes. J. biol. Chem. *255:* 1286–1295 (1980).

Kokatnur, M.G.; Kummerow, F.A.: Amino acid imbalance and cholesterol levels in chicks. J. Nutr. *75:* 319–329 (1961).

Kokatnur, M.; Rand, N.T.; Kummerow, F.A.: Effect of the energy to protein ratio on serum and carcass cholesterol levels in chicks. Circulation Res. *6:* 424–431 (1958).

Kramsch, D.M.; Aspen, A.J.; Apstein, C.S.: Suppression of experimental atherosclerosis by the CA++ antagonist lanthanum. J. clin. Invest. *65:* 967–981 (1980).

Kramsch, D.M.; Aspen, A.J.; Rozler, L.J.: Atherosclerosis: prevention by agents not affecting abnormal levels of blood lipids. Science *213:* 1511–1512 (1981).

Kritchevsky, D.; Tepper, S.A.; Czanecki, S.K.; Klurfeld, D.M.: Atherogenicity of animal and vegetable protein. Influence of the lysine arginine ratio. Atherosclerosis *41:* 429–431 (1982).

Kummerow, F.A.. Nutrition imbalance and angiotoxins as dietary risk factors in coronary heart disease. Am. J. clin. Nutr. *32:* 58 (1979a).

Kummerow, F.A.: Reply to letter by Levine and Parker. Am. J. clin. Nutr. *32:* 2169–2170 (1979b).

Kummerow, F.A.; Ueno, A.; Nishida, T.; Kokatnur, M.: Unsaturated fatty acids and plasma lipids. Am. J. clin. Nutr. *8:* 62–67 (1960).

Kummerow, F.A.; Yeh, S.J.C.; Arima, T.; Cho, B.H.S.: The value of fats in swine ration (Am. Oil Chemist's Society, Philadelphia 1974).

Kummerow, F.A.: Modification of cell membrane composition by dietary lipids and its implications for atherosclerosis; in Kummerow, Benga, Holmes, The application of specialized techniques to the study of membrane properties. (New York Academy of Sciences, New York 1983).

Lahiri, A.: Conservation of mineral resources. Commerce Annual *1976:* 47–49.

Leszczynski, D.; Toda, T.; Kummerow, F.A.: Influence of dietary sex hormones on chick lipid metabolism. Hormone metabol. Res. *14:* 183–189 (1982).

Levine, A.A.; Parker, S.J.: Letter to the Editor. Am. J. clin. Nutr. *32:* 2165–2173 (1979).

Levy, R.I.; Rifkind, B.M.; Dennis, B.H.; Ernest, N.D.: Nutrition, lipids and coronary heart disease, vol. I (Raven Press, New York 1979).

McGill, H.C., Jr.: The lesion; in Schettler, Weizel, Atherosclerosis. III. Proc. 3rd Int. Symp. on Atherosclerosis, pp. 27–38 (Springer, Berlin 1974).

McGill, H.C., Jr.: The relationship of dietary cholesterol to serum cholesterol concentration and to atherosclerosis in man. Am. J. clin. Nutr. *32:* 2664–2702 (1979).

McLaren, D.S.: The great protein fiasco. Lancet *July 13:* 93 (1974).

Mahley, R.W.; Bersot, T.P.; Innerarity, T.L.; Lipson, A.; Margolis, S.: Alterations in human high-density lipoproteins with or without increased plasma cholesterol, induced by diets high in cholesterol. Lancet *ii:* 807 (1978).

Multiple Risk Factor Intervention Trial Research Group: Multiple risk factor intervention trial: risk factor changes and mortality results. J. Am. med. Ass. *248:* 1465–1477 (1982).

National Academy of Sciences: Recommended Dietary Allowances; 9th ed. National Research Council, National Academy of Sciences (US Government Printing Office, Washington 1980).

National Diet-Heart Study Research. Group: The national diet-heart study final report. Circulation *37:* suppl. 1 (1968).

Nutrient Requirements of Swine. 8th revised edition. National Research Council. Nutritional Requirements of Domestic Animals Series, vol. 2 (National Academy of Sciences, Washington 1979).

O'Sullivan, D.: New sweeteners gain ground in Europe. Chem. Eng. News *61:* 29–30 (1983).

Proctor and Gamble Co.: Personal communication (Cincinnatti, Ohio 1968).

Rall, S.R., Jr.; Weisgraber, K.H.; Mahley, R.W.: Human apoliprotein E. J. biol. Chem. *257:* 4171–4178 (1982).

Rasmussen, H.; Matsumoto, T.; Fontaine, O.; Goodman, D.B.P.: Role of changes in membrane lipid structure in action of 1,25-dihydroxyvitamin D_3. Fed. Proc. *41:* 72–77 (1982).

Royce, S.M.; Takagi, T.; Holmes, R.; Kummerow, F.A.: Effect of dietary fatty acid isomerization and saturation on atherosclerosis in swine. Fed. Am. Soc. Biol. & Med. Als. 3022; Chicago 1983.

Schmeisser, D.D.: The influence of lysine and arginine imbalance on plasma lipids of the chick and pig (University of Illinois, Urbana 1983).

Sherrill, B.C.; Innerarity, T.L.; Mahley, R.W.: Rapid hepatic clearance of the canine lipoproteins containing only the E apoprotein by a high affinity receptor. J. biol. Chem. *255:* 1804–1807 (1980).

Simpson, J.R.: The world's beef business (Iowa State University Press, Ames 1982).

Singer, S.; Nicholson, G.: The fluid mosaic model of the structure of cell membranes. Science *175:* 720–731 (1972).

Stamler, J.: Current status of knowledge on estrogen treatment of hyperlipidemia and atherosclerotic disease; in Casdorph, Treatment of the hyperlipidemic states, pp. 310–322 (Thomas, Springfield 1971).

Sullivan, L.M.; Mathias, M.M.: Eicosanoid production in rat blood as affected by fasting and dietary fat. Prosta. Leuko. Med. *9:* 223–233 (1982).

Swern, D.; Bailey, F.: Industrial oil and fat products (Interscience, New York 1964).

Talbot, R.B.: The world food problem and US food politics and policies, 1972–1976 (Iowa State University Press, Ames 1977).

Talbot, R.B.: The world food problem and US food politics and policies, 1977 (Iowa State University Press, Ames 1978).

Taura, S.; Taura, M.; Tokuyasu, K.; Kamio, A.; Kummerow, F.A.: Ultrastructure of human thoracic aorta obtained at elective coronary bypass surgery. Artery *3:* 529–541 (1977).

Taura, S.; Taura, M.; Imai, H.; Kummerow, F.A.; Tokuyasu, K.; Cho, S.B.H.: Ultrastructure of cardiovascular lesions induced by hypervitaminosis D and its withdrawal. Arterial Wall *4:* 245–259 (1979).

Toda, T.; Leszczynski, D.; Kummerow, F.A.: Angiotoxic effects of dietary 7-ketocholesterol in chick aorta. Arterial Wall *7:* 167–175 (1982).

Toda, T.; Leszczynski, D.; McGibbon, W.H.; Kummerow, F.A.: Coronary arterial lesions in sexually mature non-layers, layers, and roosters. Virchows Arch. Abt. A Path. Anat. Histol. *388:* 123 (1980).

Tokuyasu, K.; Imai, H.; Taura, S.; Cho, B.H.S.; Kummerow, F.A.: Aortic lesions in non-laying hens with endogenous hyperlipidemia. Archs Pathol. Lab. Med. *104:* 41–45 (1980).

Toward Healthful Diets (Food and Nutrition Board of the National Research Council, National Academy of Sciences, Washington 1980).

Vital Statistics of the United States (United States Public Health Service, 1941).

Vital Statistics of the United States (United States Public Health Service, 1944).

Walker, B.L.; Kummerow, F.A.: Erythrocyte fatty acids composition and apparent permeability to non-electrolytes. Proc. Soc. exp. Biol. Med. *115:* 1099–1103 (1964).

Wallach, D.F.H.: The role of plasma membrane in disease process in biological membranes; in Chapman, Wallach, Biological membranes, vol. 2, pp. 253–293 (Academic Press, New York 1973).

Watt, B.K.; Merrill, A.L.: Handbook of the nutritional contents of food (Dover, New York 1975).

World Health Statistics Annual (WHO, Genève 1980).

Yamanaka, W.K.; Clemens, G.W.; Hutchinson, M.L.: Essential fatty acid deficiency in humans. Prog. Lipid Res. *19:* 187–215 (1980).

1980 Yearbook of International Trade Statistics, vol. 2 (United Nations, New York 1981).

F.A. Kummerow, PhD, Burnsides Research Laboratory, University of Illinois, Urbana, IL 61801 (USA)

Wld Rev. Nutr. Diet., vol. 45, pp. 42–67 (Karger, Basel 1985)

The Nutritional Status of Preschool Children in Egypt

G. Richard Jansen

Department of Food Science and Human Nutrition, Colorado State University, Fort Collins, Colo., USA

Contents

Introduction

Two of the most important public health problems in many developing countries are high rates of infant and preschool mortality and poor growth of children [31, 32]. These also are important and longstanding problems in the Arab Middle East, including Egypt [34, 35]. As will be discussed below, the causes involve, to a major extent, the interaction of poor sanitation with inadequate weaning practices. In primary health care, adequate nutrition along with oral rehydration therapy is the cornerstone of the rehabilitation of malnourished infants [22]. Even more important, in the long run, is the role of good nutrition, breast-feeding, adequate home sanitary practices, and appropriate weaning practices as part of a preventive approach to reducing mortality and improving health.

Indicators of Preschool Malnutrition in Egypt

In common with most other developing countries, the major indicators of preschool malnutrition in Egypt include anthropometric measures (height, weight), clinical and biochemical measurements and data on mortality and morbidity. Dietary data provide additional useful information. An annotated bibliography on nutrition and weaning practices in Egypt is available [29].

Anthropometric

Abdou et al. [5] in 1956–1957, carried out a nutritional status survey of 1,143 infants and children below 2 years of age in four MCH centers in Cairo. The authors reported that birth weights were above Western standards but the growth curve (height or weight) of the average child studied began to show signs of retardation as early as the third month of life. At the end of the first 2 years of life, the average child studied weighed 7 lb less and measured 12 cm less than the Iowa growth standard.

A similar study was carried out in 1965–1966 in Beheira Governorate also by *Abdou* et al. [11], wherein a total of 1,584 children aged 1–6, in the Capital City, Damanhour, and in 12 nearby villages representing agricultural, coastal, and reclamation types were studied. The study also included all seasons of the year. The extent of retardation increased with age in all groups studied. In terms of weight for age, by the end of the second year of life, 70% of the children weighed less than 90% of standard. Girls were consistently lighter than boys but there was not any significant urban-rural difference, nor did the infants brought to MCH centers differ significantly from other infants. The weight deficit became less marked after the second year, but it is likely that some of the infants with the greatest weight deficits had died by this time. During the first 6 months of life, length deficits were greater in rural than in urban areas. The peak in growth retardation was reached during the second year, and at 24 months of age 60–70% of the children not attending MCH centers were less than 90% of the height/age standard. Children attending MCH centers, especially girls, showed even greater height deficits at this point.

Shukry et al. [37, 38] reported on the incidence of protein calorie malnutrition in rural Egypt near Cairo [37] and in Cairo itself [38]. Using weight/age as the yardstick, and consistent with the earlier studies, very high rates of growth retardation were observed. Considering the two studies combined, in the 6- to 12-month age group, 68–88% of the infants were diagnosed as mild to moderate and severely malnourished combined, with again the girls showing higher rates of malnutrition than boys.

Kamel [28] reported on the results of a longitudinal study of 81 children visiting 2 MCH centers in Giza and Ein El Sira. In the age groups 0–3, 4–6, 7–11, and less than 24 months, 67, 48, 75, and 83% of the infants were considered to be malnourished (i.e. less than 90% weight/age standard). *Ali* et al. [12] reported similarly high incidences of malnutrition in infants and preschool children in the Edfan region.

A number of studies have confirmed that these retardations in linear growth that start in infancy are not necessarily overcome later in life, since significant numbers of school children and adults continue to be retarded in linear growth. *Abdou* et al. [4, 6] reported that height/age for both sexes of school-age children in the cities of Cairo, Asyut, and Aswan indicated growth retardation one standard deviation below both the Iowa standard and the growth of well-to-do children in Cairo. In rural areas of Asyut and Aswan, growth retardation was an additional standard deviation below growth in the cities. Similar results were reported by *Abdou* et al. [3] for rural and urban areas of Beheira Governorate. School children in the Sinai and in oases in the New Valley were even more growth-retarded than children in Cairo [2, 7].

In a study of 90 school boys in Sindion, *Abdou and Moussa* [9] reported that the diet was inadequate in many nutrients and the authors concluded that the growth retardation observed was environmentally, rather than genetically, determined. *Fahmi* [23] reported similar degrees of growth retardation in primary school children in Alexandria. The growth retardation observed in school children extends into adulthood. *Abdou* et al. [10] reported that adults in rural areas of Beheira Governorate were significantly more growth retarded than adults in urban areas.

During the last 5 years, three national nutritional surveys involving preschool children have been carried out in Egypt [13, 14, 21]. The MIT-Cairo University Health Care Delivery Systems Project carried out a 'weighing exercise' in 1978 [21]. In this and the two national nutrition surveys carried out with the assistance of the US Communicable Disease Center (CDC) in 1978 and 1980 [13, 14], growth retardation was expressed in two ways. In the Gomez system, weight for age is the parameter used with 75–90, 60–75, and less than 60% of the standard considered to be 1° (mild), 2° (moderate), and 3° (severe) malnutrition, respectively. The weakness of this system is that it does not distinguish between growth-stunting and wasting. The Waterlow system, used in the two 1978 surveys, defines growth stunting (chronic malnutrition) as height/age less than 90% of the standard and wasting (acute malnutrition) as weight/height less than 80% of the standard. In the 1980 CDC-assisted survey, wasting was defined as weight/height less than 85% of the standard. In all three survey to be considered, the standards used were those of the National Center for Health Statistics (NCHS) in the USA.

The MIT-Cairo University study was an evaluation of primary health care in rural areas of Egypt [21]. In the weighing exercise anthropometric data were collected for children attending MCH centers in 17 governorates in Egypt. One center in each governorate was randomly picked for evaluation out of the 132 units evaluated in other parts of the project. A total of 4,084 children, aged 0–5 years, were studied. According to the Gomez classification system in Lower Egypt 34.8, 21.0, and 7.0% were considered to be 1°, 2°, and 3° malnourished compared to values of 38.2, 14.8, and 3.3%, respectively, in Upper Egypt. There were some striking differences in malnutrition observed in the various governorates. The lowest level of malnutrition was seen in Kafr-El-Sheik where 70.3% of the children were considered normal and only 8.0% were 2° and 3° malnourished combined. In contrast, in Sharkia only 3.9% of the children were found to be of normal weight/age and 76.0% were 2° and 3° malnourished. It needs to be stated, however, that these latter data are for one MCH center in each governorate and may not be representative for these governorates. The corresponding data from this study when expressed using the Waterlow system confirmed the earlier studies [5, 11] that the primary problem in Egypt, as in most developing countries, is growth stunting which reached incidences of 33.6% in Upper Egypt and 42.7% in Lower Egypt.

In 1978 the Nutrition Institute of Egypt, with the assistance of the CDC, carried out a national nutrition survey of Egypt [13]. In this study, the frontier governorates were excluded because of their low population. 11 universes were constructed from the remaining 21 governorates using geopolitical and population criteria. Sampling procedure used a two-stage method. In the first stage, 30 sample sites were selected in each universe. In the second stage, a single household in each site was selected at random, and then data were collected on 30 children, aged 6–71 months, in a surround-

Table I. Percentage distribution of preschool children by Gomez classes and geographic area: Egypt 1978[1]

Geographic area	Degree of undernutrition, %				
	3° <60.0	2° 60.0–74.9	1° 75.0–89.9	normal 90.0+	total
Lower rural	0.6	7.8	37.4	54.2	100.0 (3,552)
Upper rural	1.0	11.9	41.9	45.2	100.0 (1,784)
Large villages	1.2	8.5	42.4	47.8	100.0 (889)
Small towns	0.7	6.0	35.5	57.8	100.0 (894)
Small cities	0.2	3.6	32.6	63.7	100.0 (897)
Total representative sample	0.8%	8.0%	38.5%	52.7%	100.0 (8,016)
Major cities and special group					
Cairo-Giza	0.7	8.4	41.7	49.2	100.0 (890)
Alexandria	0.5	4.3	38.3	57.0	100.0 (888)
Special group	–	0.5	14.0	85.6	100.0 (1,883)

[1] Data taken from reference [13]. The numbers of children surveyed are given in parentheses.

ing cluster of households. Thus, this technique is somewhat neighborhood-dependent. However, with 30 sites in each universe, any neighborhood biases should disappear or become greatly attenuated. Comparison was made with the NCHS growth norms and with data obtained from a 'special group' of 1,883 children, aged 6–71 months, obtained from private nursery schools and socioeconomically advanced families in Cairo. The 1978 survey took place from January to April, and involved a total of 9,794 children.

Data showing the distribution of preschool malnutrition in Egypt by Gomez class and according to geographic area are summarized in table I. Outside the special group, malnutrition was seen least in small cities and to the greatest extent in rural areas in Upper Egypt. However, these differences were small and the incidences of 1°, 2°, and 3° malnutrition in the total representative sample were 38.5, 8.0, and 0.8%. That the incidences of malnutrition in this survey are lower than observed in the MIT-Cairo University study is not surprising since the latter study was of children attend-

Table II. Percentage distribution of preschool children by Waterlow classes and geographic area: Egypt 1978[1]

Geographic area	Waterlow class, %				
	wasting	stunting	wasting and stunting	normal	total
Lower rural	0.5	21.5	0.3	77.8	100.0 (3,552)
Upper rural	0.2	27.0	0.6	72.2	100.0 (1,784)
Large villages	–	24.0	0.3	75.7	100.0 (889)
Small towns	0.3	14.5	0.2	84.9	100.0 (894)
Small cities	0.3	10.6	–	89.1	100.0 (897)
Total representative sample	0.3	20.8	0.3	78.6	100.0 (8,016)
Major cities and special group					
Cairo-Giza	0.6	18.8	0.2	80.4	100.0 (890)
Alexandria	0.1	15.7	0.1	84.1	100.0 (888)
Special group	0.1	1.1	–	98.8	100.0 (1,883)

[1] Data taken from reference [13]. The numbers of children surveyed are given in parentheses.

ing MCH centers whereas the CDC study was designed to be representative of the entire population.

Data for malnutrition in Egypt according to geographic area and according to Waterlow class are summarized in table II. In the total representative sample, the incidences of stunting, wasting, and stunting-wasting combined were 20.8, 0.3, and 0.3%, respectively, compared to 1.1, 0.1, and 0.0%, respectively, in the special group. Corresponding data for Cairo-Giza and Alexandria were 18.8, 0.6, and 0.2%, and 15.7, 0.1, and 0.1%, respectively.

The distribution of malnutrition according to Waterlow class and as a function of age is shown in table III. The incidence of stunting reaches a peak incidence in the 12- to 36-month-old children (26%) and then drops off somewhat. As pointed out previously, the effect of mortality on these figures needs to be considered. The incidences of wasting and stunting-wasting combined were quite low and relatively independent of age. Data for stunting and wasting as functions of age and sex are summarized in table

Table III. Percentage distribution of preschool children by Waterlow classes and age: Egypt 1978[1]

Age months	Waterlow class, %				
	normal	wasting	stunting	wasting and stunting	total
6–11	89.2	0.6	9.3	0.9	100.0 (809)
12–23	73.2	0.5	25.5	0.7	100.0 (1,816)
24–35	73.4	0.1	26.3	0.2	100.0 (1,657)
36–47	77.2	0.2	22.6	–	100.0 (1,422)
48–59	83.5	0.2	16.3	–	100.0 (1,267)
60–71	83.8	0.2	16.0	0	100.0 (1,045)
Total	78.6	0.3	20.8	0.3	100.0 (8,016)

[1] Data taken from reference [13]. The numbers of children surveyed are given in parentheses.

VII. Differences associated with sex of the infant or preschool child were minimal.

The above-described study was carried out in the winter months at a time when the incidence of diarrhea is known to be less than in the summer and when the major sickness experienced by preschool children is respiratory disease. Therefore, in 1980 a follow-up study (nutrition status survey II) was carried out in August and September by the same groups involved in the 1978 study [14]. In the follow-up study, two rural universes were selected for reinvestigation. Universe 1 was from Lower Egypt and included Damietta and Kafr-El-Sheik Governorates. The universe chosen from Upper Egypt was No. 5 consisting of El Giza, El Fayoum, Beni Suief, and El Minia Governorates. These universes were chosen for logistical, not statistical, reasons. In the follow-up study, similar samplings were followed as in the original 1978 study.

The incidences of stunting and wasting by age in the two universes and in the two surveys are summarized in table IV. In universe 1 (Lower Egypt), the differences between the two studies are not large. Stunting was 16% in 1978 and 13% in 1980, wasting 1 and 3% and stunting-wasting combined were 0 and 1% in 1978 and 1980, respectively. In universe 5, from 1978 to 1980 stunting declined from 31 to 22%, wasting increased from 1 to 7% and

Table IV. Percentage distribution of preschool children by Waterlow class, age group and universe: Egypt, 1978 and 1980[1]

Age months	Waterlow class, %								Total number examined	
	normal		wasting only		stunting only		wasting and stunting			
	1978	1980	1978	1980	1978	1980	1978	1980	1978	1980
Universe 1 (Lower Egypt)										
6–11	93	88	3	5	4	6	–	1	73	127
12–23	78	68	1	8	21	21	0	2	201	201
24–35	82	89	–	2	18	8	–	–	179	178
36–47	77	79	–	1	23	19	–	–	164	146
48–59	92	90	–	–	8	10	–	–	142	147
60–71	88	94	2	1	10	4	–	1	121	96
Total	83	83	1	3	16	13	0	1	880	895
Universe 5 (Upper Egypt)										
6–11	83	59	–	22	14	12	2	6	90	108
12–23	56	54	3	12	36	24	5	10	234	201
24–35	61	70	–	3	36	22	3	6	186	185
36–47	68	75	–	–	32	23	–	1	167	163
48–59	68	75	–	4	32	21	–	–	114	132
60–71	76	73	–	–	24	27	–	–	101	99
Total	66	67	1	7	31	22	2	5	892	888

[1] Data taken from reference [14].

stunting-wasting combined from 2 to 5%. As will be discussed later, more diarrheal disease was in fact observed in the 1980 summer study than the 1978 winter study. Interestingly, in both universes the percentages of children considered normal from a growth standpoint were nearly identical in the two studies with more wasting and less stunting observed in 1980 than in 1978.

The anthropometric data in the 1980 survey were also expressed according to the Gomez system. In the 12- to 35-month age group, the time when the incidence of malnutrition peaked, 2° and 3° malnutrition combined was 14 and 26% in universes 1 and 5, respectively. The corresponding values in the 1978 study were 8 and 20% in universes 1 and 5, respectively. The increase in moderate to severe malnutrition in 1980, as compared to 1978, is related to the increased wasting observed and is quite possibly associated with the increased incidence of diarrhea.

Clinical and Biochemical Data

In the study carried out in MCH Centers in Cairo by *Abdou* et al. [5], anemia and rickets were the major nutritional deficiency diseases observed. The incidence of clinical rickets was found to be 13%. Anemia (i.e. Hb less than 10 g/100 ml) was found in 90, 70, and 80% of pregnant women, nursing women and children under 2 years of age, respectively. The anemia in both mothers and children was 95% of the hypochromic microcytic (i.e. iron deficiency) type, and only 5% of the megaloblastic (i.e. folic acid deficiency) type. In a study carried out in 1965–1966 in Beheira Governorate, *Abdou* et al. [11] reported rachitic signs in 12.5 and 14.3% of rural and urban 0- to 2-year-olds and 7.8 and 8.7% of rural and urban 2- to 6-year-olds, respectively. Considering anemia as less than 11 g Hb/100 ml, 75% of the 0- to 2-year-olds and 68% of the 2- to 6-year-olds were classified as anemic with no significant urban-rural difference.

In 1974, *Abdel-Fattah* et al. [1] reported the results of an epidemiologic study of iron deficiency anemia in infancy. Hemoglobin concentrations less than 11 g/100 ml were observed in 90% of 4- to 6-month-old infants. The incidence of low hemoglobin levels, defined in this way, was higher for boys than girls, for fourth or higher birth orders than the first through third births, for artificially fed than for totally breast-fed infants, and for infants of anemic mothers than for infants of non-anemic mothers. *Shaheen* [36] also observed more anemia in infants born of anemic than non-anemic mothers. In his study, by 10 months of age 50.0 and 62.5% of the infants of non-anemic and anemic mothers, respectively, were diagnosed as being anemic (i.e. Hb less than 11 g/100 ml).

In the 1978 Nutrition Status Survey [13], clinical signs of kwashiorkor, and vitamin A, riboflavin and vitamin D deficiency were looked for. Virtually no clinical signs of kwashiorkor (bipedal edema) or vitamin A deficiency (bitot's spots, corneal scars, and night blindness) were seen. Angular stomatitis (relatively non-specific for riboflavin deficiency) was only detected in 2.9% of the total sample, and the highest incidence observed was 5.0% in the 48- to 71-month age group. Six commonly recognized signs of vitamin D deficiency were looked for. Virtually no children at any age group exhibited three or more of these clinical signs of vitamin D deficiency. Considering the total representative sample, 5.0% showed at least one

Table V. Percentage distribution of preschool children by hemoglobin values and geographic area: Egypt 1978[1]

Geographic area	Hemoglobin value (g/100 ml)[2], %			
	<9.5	<11.0	11.0+	total
Lower rural	14.9	44.6	55.4	100.0 (715)
Upper rural	16.5	43.4	56.6	100.0 (358)
Large villages	11.9	39.2	60.8	100.0 (176)
Small towns	8.3	30.5	69.4	100.0 (180)
Small cities	3.3	23.3	76.7	100.0 (180)
Total representative sample	12.2	38.4	61.6	100.0 (1,609)
Major cities and special group				
Cairo-Giza	5.6	35.5	64.4	100.0 (177)
Alexandria	11.8	42.1	57.9	100.0 (178)
Special group	1.9	16.9	83.0	100.0 (359)

[1] Data taken from reference [13]. The numbers of children surveyed are given in parentheses.
[2] Hemoglobin value less than 9.5 g/100 ml is indicative of severe anemia. Hemoglobin value less than 11 g/100 ml is defined by WHO as indicative of anemia.

clinical sign of vitamin D deficiency, with the highest incidence observed in the 24- to 35-month age group (8.4%).

In contrast to these vitamin deficiencies and confirming earlier work, the incidence of anemia (Hb less than 11 g/100 ml) was high in the 1978 national survey, as illustrated by the data summarized in table V. For the total representative sample 38.4% of the children were anemic with 12.2% severely anemic. The incidence of anemia ranged from 23.3% in small cities to 44.0% in rural Egypt. The incidence of severe anemia ranged from 3.3% in small cities to 15.7% in rural Egypt. Interestingly 16.9% of the special group was considered anemic by the standard used. As shown in table VI, the peak incidence of anemia (59.4%) was observed in the 12- to 23-month age group. The incidence of anemia was higher in stunted (52.9%) than in normal size (34.1%) children.

In the 1980 follow-up nutrition status survey [14], the incidence of anemia was compared with the incidence observed in the 1978 study for universes 1 and 5 as previously defined. As shown in table VII, the incidences of anemia in the two studies were closely similar, which lends addi-

Table VI. Mean hemoglobin values and prevalence of anemia among preschool children by age: Egypt 1978[1]

Age, months	Mean hemoglobin ($\pm$ SD) g/100 ml	Percent anemic Hbg < 11 g/100 ml)	Total number examined
6–11	10.7 ± 1.4	57.3	169
12–23	10.4 ± 1.4	59.4	383
24–35	11.1 ± 1.5	41.1	338
36–47	11.5 ± 1.4	31.9	270
48–59	12.1 ± 1.4	16.6	234
60–71	12.2 ± 1.2	12.8	215
Total	11.2 ± 1.6	38.4	1,609

[1] Data taken from reference [13].

tional reliability to the results of the much larger 1978 survey. In universe 1, 60 and 26%, respectively, of 6- to 23- and 24- to 71-month-old children were anemic. The corresponding data for universe 5 were 70 and 42%.

Mortality and Morbidity

Data on infant mortality rates (IMR) in Egypt by governorates were summarized by the MIT-Cairo University rural health project [25]. Differences by governorate and between urban and rural areas were observed. In 1972 the rural and urban IMRs for Egypt were 103 and 133/1,000 live births, respectively, with a weighted mean for Egypt as a whole of 116/1,000 live births. By governorate, the lowest rates were seen in Port Said and Sinai (49 and 57, respectively) and highest rates were observed for Cairo and the Red Sea area (152 and 190, respectively).

In the MIT-Cairo University project on rural health delivery systems, it was observed that the IMR for a geographic area varied inversely with size of the population covered [26]. This leads to the conclusion that under-reporting of mortality becomes more significant as size of the health unit increases and supervisory capability decreases. Therefore, the official mortality figures should be considered minimum estimates.

The results of a study on infant mortality in a single Egyptian community have been reported by *El-Fattah and El Rafie* [19]. The study was carried out in Tamouth, a small village in Giza Governorate with a population of 7,000–9,000 inhabitants. From village records, the IMR was 132–

Table VII. Mean hemoglobin (g/100 ml) values and prevalence of anemia in preschool children by age group and universe: Egypt, 1978 and 1980[1]

Age months	Mean hemoglobin (SD)		Percent anemic		Total number examined	
	1978	1980	1978	1980	1978	1980
Universe 1 (Lower Egypt)						
6–23	10.5 ± 1.5	10.7 ± 1.5	65	56	54	72
24–71	11.8 ± 1.4	11.5 ± 1.4	25	27	122	104
Total	11.4 ± 1.5	11.2 ± 1.5	37	39	176	176
Universe 5 (Upper Egypt)						
6–23	10.0 ± 1.4	10.1 ± 1.5	74	66	65	53
24–71	11.1 ± 1.4	11.1 ± 1.5	43	41	114	122
Total	10.7 ± 1.5	10.8 ± 1.5	54	49	179	175

[1] Data taken from reference [14].

144 in the period 1972–1976. During this time there were approximately 3 times as many infant deaths as deaths of preschool children. Infant mortality was fairly evenly distributed throughout the year, and was not consistently higher in summer months than winter months.

The major illnesses of children leading to mortality are known to be diarrhea, especially in summer months, and respiratory disease, especially in the winter. Considerable data on the incidences of illness in infants and preschool children were obtained in the two national surveys, and these reports need to be consulted for the details [13, 14]. The data are summarized in table VIII.

It is not possible to determine from these data whether the same or different children are involved with the different categories of illness. In general the incidence of illness declines with age, but during any single week a small child in Egypt has apparently somewhere between a 15–30% or more chance of being sick. As expected, the incidence of diarrhea was significantly higher in the 1980 study carried out in the summer months than in the 1978 study carried out in the winter, and weight/height was significantly lower in children with a recent episode of diarrhea than in children not so affected. Additional information on the epidemiology of diarrhea

Table VIII. Morbidity in the 1978 and 1980 National Nutrition Surveys[1]

Type of illness	Age months	Percent sick in a 7-day period				
		total 1978 sample	universe 1[2]		universe 5[3]	
			1978	1980	1978	1980
General sickness	6–11	21.7	23	20	20	23
	12–23	20.5	19	20	13	24
	24–35	16.9	15	18	12	25
Diarrhea	6–11	20.0	17	32	19	29
	12–23	14.0	14	22	14	26
	24–35	10.6	9	15	12	23
Fever	6–11	21.1	17	19	22	28
	12–23	20.1	18	18	16	24
	24–35	17.0	14	20	13	21

[1] Data taken from references [13, 14].
[2] Lower Egypt.
[3] Upper Egypt.

among children under 2 years of age has recently been provided by *Aziz Ali and El Geneidy* [16]. The implications of these high morbidity rates for the need to provide the nutritional support for catch-up growth are discussed later.

Dietary Data

National food consumption data that would make possible the calculation of nutrient intakes by preschool children in Egypt are not available. The 1978 national survey [13] did obtain food frequency data for the age group 0.6 years and these will next be considered. The issue of foods introduced into the infant or preschool child's diet specifically during the weaning period will be considered later.

Waslien et al. [39] reported that food energy in the diet of 49 families in Cairo with children suffering from malnutrition was supplied by food groups as follows: bread and cereals, 58%; beans, 6.3%; meat, fish, poultry, 1–5%; eggs, 0.2%; milk and cheese, 8.1%; fats and oils, 10.0%; fruits and vegetables, 4.9%; and sugar, 5%. This study is, of course, severely limited in scope; but it does provide some insights concerning the dietary pattern of some low-income families near Cairo.

National food availability data [24] indicate that nutrition problems in Egypt are more likely to be related to distribution of food in the family, and according to geographic and economic variables, than to a lack of sufficient variety and amount of food available in the country as a whole. The problem becomes particularly acute for weaning-age children since all foods in the home may not be appropriate for this age group.

In the 1978 national survey [13] dietary data were presented as units of foods from nine food groups consumed in a single day by children ages 0–6. In the report, these units are not defined but are said to be roughly equivalent to one another. The coefficients of variation for these data were quite large ranging from a low of 63% for the grain group to a high of 400% for the weaning food group. The number of food units consumed per day varied from a high of 6.41 for grain to a low of 0.18 for the weaning food group. The second most commonly consumed food was milk with 3.35 U/day.

Unfortunately, data are not available specifically for the 0- to 2-year-old child, and so the low level and high variability in the consumption of the weaning food group is in part a function of calculating the data for an inappropriate range. However, both the low level and high variability of consumption of this food group suggest that low availability and use of low-cost weaning foods with high nutritional value may be a significant contributor to nutritional problems during the weaning period.

Causes of Malnutrition in Egypt

There are many causes of malnutrition in Egypt, as in other developing countries. As will be discussed later, breast-feeding practices appear to be generally good in Egypt, and only around 5% of infants are completely weaned by 1 year of age. However, growth starts to falter by at least the sixth month of age, perhaps as early as the third month, many children die in the first year of life, and the incidence of anemia remains very high during all the preschool years. This emphasizes the importance of the weaning period.

From the available data it is not possible to say how much of the problem during weaning is caused by milk secretion rates below optimal, inappropriate weaning practices, late introduction of sufficient quantities of supplemental or weaning foods, lack of knowledge in how to prepare appropriate weaning foods, lack of availability of weaning foods or weaning food ingredients, poor sanitary practices in the home and high rates of exposure to respiratory and diarrheal disease, as well as parasitic infection. All are probably involved. In the case of nutritional anemia, it is not clear to what extent the problem is caused by inadequate consumption of available dietary iron, or high rates of parasitic infection. Again both are involved. In any case, additional sources of available iron in the diet would be desirable.

A likely scenario in the development of protein energy malnutrition is the following. The infants are born of normal birth weight and lactation is generally satisfactory for the first 4 and in some cases 6 months of life. Beyond this point the supply of breast milk becomes increasingly inadequate. In many cases the mother, either through ignorance, cultural preferences, lack of money or lack of suitable foods does not feed the child enough supplemental food of appropriate nutritional quality, including calories, protein, iron, zinc, and other nutrients, to meet nutritional needs. The infant is exposed to an environment where sanitary practices and waste disposal are not satisfactory. Infections are endemic. As a result, as partial weaning starts and even when breast-feeding is the only source of food, the infant and then the preschool child suffer repeated bouts of illness. During these bouts appetite is depressed and nutrient loss increased. In addition, the mother may mistakenly withhold fluids and nutrients, including breast milk. Oral rehydration can restore fluid balance but a generous nutrient intake is needed to allow catch up growth to occur. What seems to be more typical is that growth never catches up to the genetic potential. As a result, after repeated bouts of sickness, the infant or child either dies, suffers acute malnutrition, or exhibits growth stunting and nutritional anemia [15].

The respective roles of breast-feeding and weaning practices on the one hand and home sanitary and waste disposal practices on the other in causing infant mortality, wasting and growth stunting, are impossible to separate. Both are undoubtedly involved in a vicious cycle leading to malnutrition or death. Without in any way underestimating the importance of safe water and good sanitary practices in both the home and village, which may well be the dominant factors, there would appear to be opportunity for improvements in weaning practices and the availability of nutritious weaning foods with good acceptability and low cost to reduce the incidence of malnutrition in Egypt. However, in order for significant improvements in growth and reductions in infant mortality to be made, clearly educational programs will need to emphasize breast feeding, good sanitary practices as well as the proper use of nutritious weaning foods.

Table IX. Percent of sample children breast-feeding only, breast-feeding with food supplement, and not breast-feeding[1]

Food source	Age groups, %		
	6–11 months	12–23 months	24–35 months
Breast-feeding only	23.8	3.3	0.1
Breast-feeding with food supplement	68.0	62.6	9.9
Not breast-feeding	8.2	34.1	90.0
Total	100.0	100.0	100.0
(number of children)	(781)	(1,749)	(1,572)

[1] Data taken from reference [13].

The Weaning Period in Egypt

The above considerations strongly suggest that the weaning period, weaning practices and the availability of weaning foods, represent an area where improvements might make some contribution to reducing the incidence of infant and preschool malnutrition in Egypt. Breast-feeding and weaning practices both are important in this context.

Breast-Feeding Practices

Nasser et al. [33] evaluated the breast-feeding practices of 200 mothers with infants less than 6 months of age and who were attending MCH centers in El Darb El Ahmar. Considering the entire sample of 200 infants, 117 were totally breast-fed, 48 were breast-fed with some artificial milk, and 35 were artificially fed only. The nature of the artificial milk formula was not described.

The 1978 and 1980 National Nutrition Surveys provide considerable information on breast-feeding practices in Egypt, and the ages at which partial and complete weaning occur. For the first 6 months of life, the proportion of infants still being breast-fed ranged from a high of 96% in rural areas of Upper and Lower Egypt, and large villages to a low of 85–90% in small towns, small cities, Cairo-Giza and Alexandria [13]. Additional information on the ages of partial and complete weaning are summarized in table IX. During the second 6 months of life, 68.0% of the infants received food supplements and only 8% were not breast-fed. During the second year

Table X. Prevalence of chronic undernutrition in preschool children 24–71 months of age as related to age at which weaning was completed: Egypt 1978[1]

Age weaning completed months	Age in months of survey child, %			
	24–35	36–47	48–59	60–71
<3	24.0 (59)	9.6 (46)	14.5 (42)	11.0 (27)
3–5	7.0 (18)	8.7 (19)	5.5 (16)	13.8 (12)
6–11	30.8 (92)	17.7 (66)	9.5 (54)	9.9 (54)
12–17	28.0 (363)	22.6 (255)	14.7 (239)	12.3 (185)
18–23	22.0 (503)	21.2 (395)	15.9 (343)	15.4 (270)
24+	27.8 (381)	25.8 (518)	18.7 (489)	19.6 (421)
Total (number of children)	(1,416)	(1,299)	(1,183)	(969)

[1] Data taken from reference [13]. The numbers of children surveyed are given in parentheses.

of life, 63% of the infants were breast-fed with a food supplement and 34% were totally weaned. During the third year of life, 90% of the infants were totally weaned, but 10% were still being partially breast-fed.

The prevalence of chronic undernutrition as a function of the age at which weaning was completed is summarized in table X. The lowest incidence of malnutrition was observed in children totally weaned during the third to fifth months of life, and the highest incidence in children still being breast-fed beyond the second year of life. The data are uncorrected for important demographic and economic variables, however. It is likely, for example, that infants weaned during the third to fifth months of life are from the higher socioeconomic groups and those still being breast-fed during the third year from lower socioeconomic groups.

The incidences of diarrhea and malnutrition in 6- to 11-month-old infants in relation to breast-feeding practices in the 1978 and 1980 surveys [13, 14] are summarized in table XI. In the 1980 study, carried out during the summer months when diarrhea is more prevalent, there was no difference in the incidence of diarrhea in infants exclusively breast-fed than in infants receiving supplemental food. In contrast in the 1978 study carried out during the winter months, the incidence of diarrhea in exclusively breast-fed infants was twice as high as in infants receiving food supplements. This suggests that introducing food supplements was not associated

Table XI. Prevalances of acute undernutrition, combined second- and third-degree Gomez classes, and diarrheal illness for preschool children aged 6–11 months by breast-feeding status and by universe: Egypt, 1978 and 1980[1]

| | 1978 | | | | 1980 | | | |
| | breast-feeding only | | other[2] | | breast-feeding only | | other | |
	percent	number	percent	number	percent	number	percent	number
Universe 1 (Lower Egypt)								
Acute undernutrition	4	48	–	20	5	63	7	61
Combined second- and third-degree Gomez classes	10	48	10	20	8	63	20	61
Diarrhea	21	48	10	20	38	63	26	61
Universe 5 (Upper Egypt)								
Acute undernutrition	2	66	4	23	36	50	15	41
Combined second- and third-degree Gomez classes	20	66	9	23	52	50	34	41
Diarrhea	23	66	9	22	30	50	27	41

[1] Data taken from reference [14].
[2] 'Other' refers to breast-feeding with a food supplement, or not breast-feeding.

with increased rates of diarrhea. However, it is possible that the exclusively or partially breast-fed groups differed in other important socioeconomic or cultural variables. In any case, no consistent pattern of incidence in 2° and 3° malnutrition was apparent as a function of breast-feeding and food supplementation practices.

Weaning Practices

More information is available on breast-feeding than on weaning practices, particularly information pertaining to the types, amounts, nutritional value and time of introduction of food supplements given to infants and children during the weaning period. *Morgan and Adelman* [30] reviewed the available data and listed a variety of cereals, legumes, and dairy products, including buffalo milk fed to children during the weaning period. In addition, their report summarizes several traditional weaning food mixes such as 'Mahalabya' (starch, sugar, and buffalo milk) and 'Seven Seeds' (4 cereals plus 3 legumes) that appear to be in declining use or are not used at all.

Processed weaning foods, such as Supramine and Title II Instant Corn Soy Milk (ICSM) appear by general acknowledgement to not be well accepted, but quantitative data are not available, nor are the reasons for poor acceptability well understood. A reasonable hypothesis is that part of the acceptability problem with ICSM is a lack of clear identity of the product, no clear understanding of what the product is for and how it should be used and prepared.

De Treville [18] also reviewed some of these same aspects of weaning foods in Egypt discussed by *Morgan and Adelman* [30] and provided information on the history of weaning foods in Egypt, including the development of protein-rich food mixtures at the Nutrition Institute. Although many of these mixtures are of undoubted nutritional value, as confirmed by preclinical and even clinical studies, none appear to have gotten beyond the laboratory stage.

The above reviews provide general overviews of traditional as well as centrally processed food mixtures that appear to be fed, or to have been fed, to children during the weaning period. However, they provide no specific data on the important questions of (1) at what age what specific foods are introduced, (2) what are these foods and in what amounts are they given, (3) what is the nutritional value of these foods, (4) how much breast milk of what composition is typically secreted at varying durations of lactation, and finally (5) how do the answers to the above questions vary with geographic, socioeconomic, and other demographic factors. Answers to these questions are not currently available for Egypt.

The 1978 National Survey [13] provided some information on the age of introduction of weaning foods. In infants less than 3, 3–5 and 6–11 months old, 4–8, 9–17, and 50–53%, respectively, were introduced to food supplements. The results of a study of weaning practices in urban and rural Egypt has recently been published [17]. The most commonly used weaning technique among low-income mothers was to keep the infant off the breast by constant discouragement, while for middle- and upper-income mothers gradual weaning was the rule. Special weaning foods were prepared by 59% of urban mothers and only 11% of rural mothers. The type of specially prepared weaning food in rural areas was usually a starchy food (68%). In urban areas starchy foods, mixed diet and protein-rich foods comprised 30, 40, and 19%, respectively, of specially prepared weaning foods. Low-income mothers were included to use starchy type weaning foods (64%). It was noted by these authors that in rural areas and for low-income mothers in urban areas, the child was introduced to adult foods between 12 and 24 months without reducing the number of breast-feedings. This recent study highlights the potential importance of either providing or helping the mothers prepare a high quality weaning food, especially in rural and/or low-income areas where very few mothers prepare a special weaning food for the child, and when they do where it tends to be a starchy gruel.

Weaning practices and supplementary feedings of infants and young children attending MCH centers in Cairo in 1956–1957 were studied by *Abdou* et al. [8]. The ages at which various supplements were given are summarized in table XII. Although no data are provided on the amounts of these supplements given, the authors concluded that the amounts of food given were inadequate and that the supplements were generally poor in nutritional quality, being low in protein, calcium, and iron. The importance of early and continued feeding of the weaning-age child during episodes of diarrhea has recently been emphasized by *El Geneidy and Aziz Ali* [20]. *Hussein* et al. [27] evaluated weaning practices of 100 mothers attending an MCH Center in Shoubrament, a village in Giza Governorate. Supplementary feeding of foods other than milk was initiated at the age of 6 months by 64% of the mothers, with an additional 26% initiating supplementary feeding by the age of 1 year. The introduction of protein-rich foods in the form of animal foods, such as eggs, meat, fish, or chicken, was typically delayed until the age of 18 months or later because of a belief that digestive disorders or allergies would result if such foods were given earlier.

Table XII. Weaning practices in Egypt[1]

Different food items	Age group in months and percent of children in each grouping[2]					
	0–1	1–3	3–6	6–9	9–12	12–24
1 Anise or caraway water	53.3	57.1	24	2.8	–	–
2 Fenugreek water	6.6	5.3	2	–	–	–
3 Rice water	–	7.1	8	6.1	4.3	2.7
4 External milk	–	12.5	10	17.1	18.7	22.2
5 Orange juice	–	–	2	2.8	13.0	22.2
6 Tomatoes (or juice)	–	–	–	–	4.3	16.6
7 Starch pudding (balouza)	–	–	2	2.0	2.3	2.7
8 Milk pudding (mehallabia)	–	–	2	5.7	10.5	16.6
9 Rice (boiled)	–	–	–	8.5	13.0	69.4
10 Potatoes	–	–	–	17.1	21.7	66.6
11 Biscuits	–	–	–	17.1	14.3	22.2
21 Bread	–	–	–	2.8	4.3	68.8
31 Stewed wheat (belila)	–	–	2	2.8	4.3	8.3
14 Stewed broad beans (foul medammes)	–	–	–	14.3	13.0	47.2
15 Cooked vegetables	–	–	–	2.8	4.3	37.2
16 Fruits	–	–	–	–	–	5.5
17 Eggs	–	–	–	–	–	8.3
18 Cheese	–	–	–	–	–	25.0
19 Meat	–	–	–	–	–	13.9

[1] Data taken from reference [8].
[2] The percentages of children receiving the different additions of food items in the different age groups.

In summary, the weaning period is a time of nutritional vulnerability for the infant and young child. Although many factors are involved, an important one is the failure to supplement adequately the nutrient intake of the breast-fed infant beyond the fourth month of life. Weaning foods and weaning food ingredients containing generous quantities of protein and other nutrients and fed at adequate energy levels do not appear to be widely used, whether through ignorance of their importance, their non-availability in the market, the lack of income to purchase or the lack of knowledge of how to prepare.

Discussion

From the foregoing considerations, it is clear that the infants and young children are nutritionally vulnerable during the weaning period. Under optimal conditions, well-fed mothers can generally meet the needs of their babies completely through exclusive breast-feeding for the first 6 months of life. Beyond that point supplementary feeding is needed. In many developing countries, Egypt included, growth appears to start faltering in some cases as early as the third month of life [40–42] as discussed earlier. In addition, episodes of diarrhea or respiratory disease during the weaning period are common and these cause weight loss and retard growth. It is the purpose of a weaning food to meet the nutritional needs of the breast-fed infant that are not being met by breast milk and also to ease the transition from breast milk, which is fully digestible, of high energy content and of high nutritional quality, to the family diet which, in developing countries, is often of marginal nutritional quality, low in energy, and is not nearly as digestible. In addition to the obvious need for protein in terms of weight gain and nitrogen retention, *Waterlow* [41] suggested that more protein may be needed for linear growth than for nitrogen retention. This hypothesis is by no means established, but since in Egypt, and indeed in many developing countries, growth stunting is more common than wasting, it is a reasonable possibility.

Stunting and wasting should not be considered as totally separate phenomena, nor should rehabilitation and normal feeding be divorced from one another. In view of the high incidences of diarrhea and/or respiratory disease, the normal growth pattern is one characterized by numerous bouts of illness with loss of appetite and weight loss followed by either recovery or, in too many cases, death. In the rehabilitation period quite often the dietary intake is sufficient for recovery but not for catch-up growth. The result is growth stunting in recovered children, with the incidence of wasting observed the proverbial tip of the iceberg. The observed increased in wasting in the 1980 national survey as compared to the 1978 study, supports a role of diarrhea in exacerbating wasting. As has been pointed out, these episodes of illness also are involved in causing growth stunting, and the diet of the weaning age child needs to supply generous quantities of energy and protein for catch up growth. As important as breast-feeding is, the data show clearly that by itself breast-feeding during the first year of life will neither prevent diarrhea nor malnutrition.

The above considerations indicate that even under conditions where the diet consumed by the young child is adequate in amount and nutritional quality to provide for normal growth in the absence of disease, wasting and stunting can still occur. In addition, the strong indicators are that in a significant proportion of the weaning age population, the nutrient intake beyond the fourth month of life is not adequate because of a failure to introduce supplemental foods of appropriate nutritional quality soon enough and in adequate amounts. In some cases, the failure to introduce adequate weaning foods may be a result of ignorance and in other cases inadequate income. In either case, the result is the same. Continued breast-feeding is of undoubted importance. In addition, providing an appropriate weaning food for mothers to use or helping mothers learn how to prepare a weaning food will not only provide infants and preschool children with the additional energy and nutrients that are needed but also provide an opportunity to help the mother better understand the importance of supplemental foods for her infant. The use of growth charts should facilitate the nutrition education process. It is very important that the educational programs emphasize home sanitation.

Acknowledgements

The author would like to acknowledge very helpful discussions with Drs. *Osman Galal, Mohamed Hussein, Farouk Shaheen, Amin Said*, and *Wafa Moussa Antonios* from the Nutrition Institute and Dr. *Onei Metwalli* from the National Research Council in Cairo. Travel to Egypt was made possible by funding from the Office of International Cooperation and Development, USDA and USAID.

References

1   Abdel-Fattah, M.; Shalaby, S.; El-Ashmawi, S.: An epidemiological study of iron deficiency anemia in early infancy. Gaz. Egypt Paed. Ass. *22:* 143–147 (1974).

2   Abdou, I.A.: A comparative study of heights and weights in oases of the New Valley, and Villages of the Nile Valley. Bull. Nutr. Inst. Cairo Egypt UAR *2:* 84–110.

3   Abdou, I.A.; Abdel-Azim, M.; Bishara, F.F.; Harez, S.: Heights and weights of both sexes of the age group 7–19 years in different rural urban and industrial sectors of Beheira Governorate as an indication of their nutritional status. Bull. Nutr. Inst. Cairo Egypt UAR *3:* 42–64 (1967).

4   Abdou, I.A.; Ali, H.E.; Basiouni, A.B.; Nafie, A.M.; El-Shazli, A.E.; Abdel-Kader, M.S.: Nutritional deficiencies, goitre, dental caries and parasitic infestation among school children in rural and urban areas of Asyut and Aswan Governorates. Bull. Nutr. Inst. Cairo Egypt UAR *3:* 106–123 (1967).

5   Abdou, I.A.; Ali, H.E.; Lebshtein, A.K.: A study of the nutritional status of mothers,

infants and young children attending maternal and child health centers in Cairo. I. The nutritional status of infants and young children. Bull. Nutr. Inst. Cairo Egypt UAR *1:* 9–19 (1965).

6   Abdou, I.A.; Ali, H.E.; Mahfouz, A.H.: The nutritional status of school children in rural and urban areas of Asyut and Aswan Governorates as indicated by their heights and weights in comparison to Cairo children. Bull. Nutr. Inst. Cairo Egypt UAR *3:* 88–102 (1967).

7   Abdou, I.A.; El-Kilany, M.S.; Ahmed, B.; Khalil, M.A.: Nutritional conditions in Sinai. II. The nutritional status of population groups in Sinai. Bull. Nutr. Inst. Cairo Egypt UAR *2:* 32–46 (1966).

8   Abdou, I.A.; Lebshtein, A.K.; Kassim, T.A.: A study of the nutritional status of mothers, infants and young children attending maternity and child health centers in Cairo. III. Weaning practices and supplementary feeding of infants and young children. Bull. Nutr. Inst. Cairo Egypt UAR *1:* 39–48 (1965).

9   Abdou, I.A.; Moussa, W.A.: Study of dietary factors causing growth retardation of boys in the Egyptian village. Egypt. J. Nutr. *1:* 43–58 (1975).

10   Abdou, I.A.; Shaker, E.M.; Abdel-Azim, M.; Hafez, S.; Shahin, F.: Heights and weights of adults of both sexes in different rural, urban and industrial sectors of Beheira Governorate as an indication of their nutritional status. Bull. Nutr. Inst. Cairo Egypt UAR *3:* 68–83 (1967).

11   Abdou, I.A.; Shaker, M.S.; Bishara, F.F.; El-Megharbel, M.K.: A comparative study of the nutritional status of infants and preschool children in different types of villages, urban sectors and MCH Centres of Beheira Governorate. Bull. Nutr. Inst. Cairo Egypt UAR *3:* 6–35 (1967).

12   Ali, H.E.; Mansour, A.M.; Dakroury, A.M.; Abdelhay, M.B.; Ghoneme, F.: Investigations on bladder stone disease of upper Egypt. I. The nutritional status of preschool children in Edfou. Bull. Nutr. Inst. Cairo Egypt UAR *5:* 13–25 (1975).

13   Arab Republic of Egypt. National nutrition survey, 1978 (Office of Nutrition, AID, Washington 1978).

14   Arab Republic of Egypt. National Nutrition Survey II, 1980 (Office of Nutrition, AID, Washington 1980).

15   Ashworth, A.: International differences in infant mortality and the impact of malnutrition: a review. Hum. Nutr. clin. Nutr. *36C:* 7–23 (1982).

16   Aziz Ali, F.A.; El Geneidy, M.M.: A study of diarrhea among children under 2 years of age in Alexandria. I. Epidemiology of diarrhea. Bull. High Inst. publ. Hlth *12:* 141–154 (1982).

17   Darwish, O.A.; Amine, E.K.; El-Sherbiny, A.F.; Aly, H.A.; Salama, M.H.: Weaning practices in urban and rural Egypt. UN Univ. Food Nutr. Bull. *4:* 1–6 (1982).

18   Treville, D. de: Background information essential to commencing detailed investigations of the feasibility of producing processed weaning foods, or weaning food components in Egypt. Report submitted to USAID/Cairo, 7 June 1983.

19   El-Fattah, A.; El Rafie, M.: Child (0–5 years) morbidity and mortality in an Egyptian rural community. Supplement of Ain Shams Medical Journal. Proc. 1st Ain Shams Medical Congr., 1977, vol. II, pp. 57–76.

20   El Geneidy, M.M.; Aziz Ali, F.A.: A study of diarrhea among children under 2 years in Alexandria. 2. Assessment of feeding practice among cases of diarrhea in Alexandria. Bull. High Inst. publ. Hlth *12:* 255–273 (1982).

21   El Lozy, M.; Field, J.O.; Ropes, G.; Burkhardt, R.: Childhood malnutrition in rural Egypt: results of the Ministry of Health. Weighing exercise. CO/MIT TAP Rep. 80-10, 11 June 1980.

22   El-Sherbini, A.F.; Fahmy, S.I.; Eid, E.E.; Goda, M.Y.; El-Tantarvy, A.S.; El-Sayyed, L.: The use of oral rehydration in infantile diarrhea. J. Egypt. publ. Hlth Ass. *53:* 82–104 (1978).

23   Proceedings of Workshop in Nutrition and Health in Egypt. Cairo 1979, pp. 90–95 (Office of Nutrition, AID, Washington 1979).

24   FAO Provisional Food Balance Sheets 1972–1974 Average (FAO, Rome 1977).

25   Field, J.O.; Ropes, G.: Infant mortality, birth rate and development in Egypt. CO/MIT TAP Rep. 80-8, 8 May 1980.

26   Field, J.O.; Ropes, G.: The influence of the health system on the reported incidence of infant mortality and birth weights in rural Egypt. CO/MIT TAP Rep. 80-9, 11 June 1980.

27   Hussein, M.A.; Ghazi, C.; Bassiuny, N.M.; Fayad, I.; Nelson, C.: Influence of cultural and traditional beliefs of mothers in an Egyptian village upon their nutrition habits and child feeding. Egypt. J. Nutr. *2:* 85–94 (1978).

28   Kamel, L.: Malnutrition in preschool children; in Proc. Workshop on Nutrition and Health in Egypt. Cairo 1979, pp. 84–89 (Office of Nutrition, AID, Washington 1979).

29   Mayers, M.: Annotated bibliography on nutrition and weaning practices in Egypt. Report prepared for nutrition consultants, AID/Cairo 1982.

30   Morgan, R.W.; Adelman, C.: Improving the nutritional status of weaning age children in Egypt: a review of existing projects and recommendations for new projects in conjunction with or supplementary to the present. CRS/MCH Title II Foods Program, International Food and Nutrition Program (MIT, Cambridge 1982).

31   Morley, D.: Pediatric priorities in the developing world (Butterworth, London 1973).

32   NAS/NRC Preschool Child Malnutrition: Primary deterrant to human progress. Publ. 1282 (National Academy of Sciences, National Research Council, Washington 1966).

33   Nasser, S.; El Fattah, M.A.; Emam, Z.: Maternal factors influenced feeding. Egypt. J. Nutr. *2:* 65–76 (1978).

34   Patwardhan, V.N.; Darby, W.J.: The state of nutrition in the Arab Middle East (Vanderbilt University Press, Nashville 1972).

35   Scrimshaw, N.S.; Gabr, M.: Practical approaches to combat malnutrition with special reference to mothers and children, Cairo, Egypt (Ministry of Health, Cairo 1977).

36   Shaheen, F.A.: Epidemiological study of anemias in pregnancy and their effects on the newborn in rural communities, PhD thesis, Alexandria (1979).

37   Shukry, A.S.; Barakat, M.R.; El-Ganmel, R.; Kamel, L.M.: An epidemiological study of protein-calorie malnutrition among rural population in Egypt. Gaz. Egypt Paed. Ass. *20:* 151–167 (1972).

38   Shukry, A.S.; Labib, F.M.; Kamel, L.: Assessment of the health and nutritional standards of infants and preschool children in high density Cairo community. Gaz. Egypt Paed. Ass. *21:* 47–56 (1972).

39 Waslien, C.I.; Yacoub, N.A.; Rizk, M.E.: The diet of families with children having protein-calorie malnutrition. Proc. 6th Symp. Nutr. Health Near East, pp. 402–405 (American University, Beirut 1971).
40 Waterlow, J.C.; Ashworth, A.; Griffiths, M.: Faltering in infant growth in less developed countries. Lancet *ii:* 1176–1178 (1980).
41 Waterlow, J.C.: Observations on the suckling's dilemna – a personal view. J. hum. Nutr., Lond. *35:* 85–98 (1981).
42 Waterlow, J.C.; Thomson, A.M.: Observations on the adequacy of breast-feeding. Lancet *ii:* 280–242 (1979).

G.R. Jansen, PhD, Department of Food Science and Human Nutrition,
Colorado State University, Fort Collins, CO 80523 (USA)

Wld Rev. Nutr. Diet., vol. 45, pp. 68–96 (Karger, Basel 1985)

Nutritionally Beneficial Cultural Practices

Christine S. Wilson

Department of Epidemiology and International Health, University of California,
San Francisco, Calif., USA

Contents

Introduction

Most nutrition scientists aware of how culture – a given society and its technology and beliefs – affects nutrient intakes, tend to the view that the taboos and other complex rules people oblige themselves to observe with regard to eating, have adverse effects on nutritional well-being [123]. Such comments as, 'there is no concept of foods being good for children' [44] are rather frequently found in the literature.

There are two ways to look at this situation. First, it should be recognized that statements such as those cited above are evaluations, usually made by outsiders, based on ethnocentric preconceptions of what good nutritional behavior ought to be. People may have been eating in seemingly constricted ways for generations, while surviving to procreate and maintain their kind. Their dietary practices may not be so detrimental as would at first appear, or the researchers may have overlooked or missed some nutritionally relevant acts. Secondly, the practices may indeed be less than optimal, but the people performing them may have adapted physiologically over time to attain some newer balance with respect to a nutrient or nutrients, while maintaining intact the integrity of the cultural framework in which both practices and people exist. Trying to amend the nutritional problems might disrupt the culture, making for unhappy people, deprived, for example, of their pigs, feasts and prestige, but with white teeth, as *Oliver* [89] described some Melanesians he encountered who had been persuaded by a missionary to give up meat and chewing betel nut (*Areca catechu* which, mixed with saliva and the calcium compound commonly accompanying the quid, turns lips, gums, teeth and spittle a vivid red).

However, although the published numbers of taboos or interdictions against certain foods for certain persons in certain states or statuses are preponderant over positive statements regarding food, there are some references to nutritionally advantageous practices prescribed or recommended by a culture, even in societies in which general attitudes of some of its members toward diet and eating may appear harsh. Such practices and the beliefs attending them can be cataloged in a variety of ways: according to what is done to or with the food, by the behavior of the particular subgroup of the population for whom it is recommended, or alternatively, the behaviors might be considered in terms of the nutrients they affect, or as specific sets of beliefs about qualities of foods.

Because cultural practices that influence food and eating may affect more than one nutrient for more than one geographic subgroup, the data presented here have, in several instances, been allowed to speak for themselves as to nutrients that may be involved. In such cases the phenomena are, instead, examined in terms of the ways in which foods are treated (manipulated) and with regard to social beliefs about foods, with examples that affect particular members of a population, as the data have indicated.

Much of the information presented here has been extracted from literature on anthropology. However, as the reference list indicates, nutrition scientists, clinicians and others who work in the health professions have on occasion published observations on worthy nutritional practices and behavior. As interactions grow between scientists in both these fields, nutrition and anthropology, future collaborative research may discover further instances.

Nutritionally Beneficial Manipulations of Food

Manipulations of food are here meant not only to include preparation and cooking practices that may enhance nutrient content or the availability of existing nutrients, but also additions made to the food, and the combinations or ways in which they are eaten.

Treatment of Foods – Preparation and Cooking

A simple example of a cultural food practice familiar to most western peoples is baking potatoes, usually in the skin, which conserves a considerable amount of the available ascorbic acid. Those who eat the skin as well as its contents receive further nutritional benefits that include some minerals as well as this vitamin. The handmilling (pounding to remove the husk) of grains that preserves intact much of the germ with its vitamins and minerals is another obvious example from more traditional cultures.

Cooking practices may enhance or reduce the nutritional value of a food. *Draper* [32] has pointed out that the minimal cooking of meat by aboriginal Eskimos helped preserve its antiscorbutic level of vitamin C. The traditional method used by the Flathead Indians of Montana to cook the root of the lily camas *(Camassia quamash)*, prolonged steaming in a pit lined with hot stones, has been shown to change its complex, undigestible polysaccharide, inulin, to sugars readily absorbable by human beings [65]. Oats, like other grains, have a relatively high content of phytates, which bind minerals indissolubly, such as calcium in the milk usually taken with oatmeal. In Scotland oatmeal porridge used to be simmered slowly all night on a low fire at the back of the stove. This long slow cooking considerably reduced its phytic acid content. The well-known Chinese practice of cutting food very fine before stir-frying quickly over high heat is done to save scarce fuel [2]. At least two nutritional advantages ensue. The swift cooking saves a number of nutrients from destruction common with longer cooking times, while it breaks down plant cell walls sufficiently to make digestion easier. When the foods so prepared are meats, the bones too are cut through. Upon addition of acid liquid (wine, vinegar) calcium and other minerals are leached from the bones into the cooking fluid, making them available to the consumer. French people sometimes cook fish in sorrel (sour grass, *Rumex acetosa*), which has a similar effect on the calcium in the bones. Cultures in which fish or cut-up meat is wrapped in a leaf before cooking may also get calcium from bones in this way, due to acids in the leaves [35], provided the cooking water is taken. This practice was rather common in the South Pacific, where foods were wrapped in leaves before being steamed in earth ovens, pits lined with heated stones. *Farb and Armelagos* [35] cite it for Africa. Some people customarily chew bones of meat or larger fish. *Brewer* [14] reported this practice for Somoans in Los Angeles.

The iron content of foods, particularly acidic foods, that are cooked in iron vessels is enhanced by similar interactions of the food liquid with the pots. Indeed, such iron is sufficiently readily available as to cause problems of toxicity in some African populations whose food is cooked exclusively in these containers [15].

The liquid added to foods for cooking or produced by the foods themselves in cooking contains some of the water-soluble vitamins and minerals originally present in the food. Southerners (in the Southeast of the USA) have been wont to ascribe the fine white teeth of Negro Southerners to their usual consumption of this 'pot likker' along with the vegetables or other foods. Pot liquor was less commonly taken by white Southerners.

The methods used by native South American Indians to remove from manioc (*Manihot utilissima,* also termed cassava and tapioca) the cyanogenetic glucoside that breaks down after harvest to become prussic acid are rather well known [60]. They include leaching (soaking), rotting and heating, ofttimes accompanied by grating, or squeezing in specially made baskets [81]. Less known is the fact that the Indians of the Andes still treat a bitter variety of potato (*Solanum juzepczukii*) as they did before the Spaniards came by soaking the tubers and then exposing them to freezing winter temperatures [120]. The procedure removes the glycoalkaloid causing the bitter taste, at the same time making it possible to store the potatoes for as long as 4 years. Although both manioc and potatoes are nutritionally mainly starch, they are important energy sources in large areas of the world, and as such spare other more nutritionally 'dense' foods such as animal protein from being used for energy. (The potato is, of course, a better source of vitamins and minerals than prepared manioc, and its protein content, although low, is of relatively good quality [70].) The indigenously developed processing methods make these widely available carbohydrate foods safe to eat.

Sometimes a food is eaten unprepared or uncooked. Consuming raw food may be nutritionally beneficial in some instances. *Draper* [32] has pointed out that the Eskimo practice of eating meat which had been cooked little or not at all preserved its antiscorbutic levels of vitamin C. Eskimos also ate most of the organs of the animals they killed, including the adrenal glands, which are a good source of ascorbic acid, one of the parts of the body where it is sequestered.

Additions to Foods

Items added to foodstuffs, intentionally or inadvertently, may enhance the nutritional quality of the food, or contribute beneficial nutrients present in the additive that are low or absent in the food itself. The classic example is the staple maize corn as prepared in Mexico and Central America. Soaking the kernels in an alkali solution to soften the stiff seed coat was long ago found to make it easier to pound them into meal. US scientists who analyzed the nutrients in the dough *(masa)* made from the meal found that the lime water (calcium carbonate) used as this vehicle added its calcium to the meal, making the resulting *tortillas* (cooked cakes) good sources of available calcium for a population that consumes little milk [3]. Later research by other workers showed that the lime treatment released existing niacin in the maize that was bound to a protein. The lime treatment thus made the vitamin available to the consumer [68]. This demonstration helped explain the absence of pellagra among maize corn eaters of Latin America who process the maize meal in this way [62].

Calloway et al. [20], studying the way Hopi and Papago Indians prepared their traditional staple corn foods, found that these peoples ashed fresh greene plants (salt bush or the pods of beans and their vines), adding the alkaline product to the water of the batter of the blue maize meal that was made into a pancake-like *piki* bread. (Addition of the ash maintains the blue color of the corn, the appearance and hue of which are the cook's criteria of how much ash to add.) This practice added sufficient minerals to an average day's intake of the bread to provide half the adult daily need for calcium, manganese and zinc, and all the requirements for magnesium and iron.

The Sanio-Hiowe people of the Upper Sepik region of New Guinea were found to obtain, among other nutrients, 234 mg of calcium and 69 mg of magnesium from the ash of the midrib of the sago palm, which they burned to obtain a kind of salt [114]. This alkaline ash spared nitrogen otherwise needed to neutralize the acid urine usual on the high vegetable diet they customarily consume. Peoples who do not have access to sea salt or salt deposits may get iodine and other needed minerals from plant materials similarly treated.

In several parts of the world (Asia, Ethiopia, Andean and lowland South America) narcotic plant parts are chewed along with a calcium-containing compound. In Asia, the basis of the quid is the betel nut, referred to above in the Introduction. In Ethiopia this substance is *ch'at (Catha edulis),* a leaf [106] and in South America it is the coca leaf *(Erythroxylon coca)* that is the source of cocaine [81]. In each case the calcium compound interacts with the pharmacologic stimulant to make it active. While the nutritional contributions of the calcium component to the chewers (some of it is swallowed along with the saliva produced) may be legitimately questioned, in the Andes other nonfood calcium sources are ingested with food [6]. One of these is a powdered earth called *cal* by the Quechua Indians [6], which is mixed with the staple porridge made from the local, high-protein grains named *quinua* or *canihua* in the local language, amaranths of the *Chenopodium* species [27]. The other is *cha'qo,* a clay suspension eaten with potatoes.

Simple additions to foods may have more than one benefit. Adding milk to tea contributes more than calcium to the drinker. Tea contains tannin, which is acidic. The milk protein helps break it down to prevent its being absorbed, and the calcium lowers the acidity [35].

Fermentation

It has long been known that the process of fermentation results in increases in beneficial nutrients in the product. It may be commercially undertaken for that purpose. Fermentation is an ancient process, probably arrived at by observation of examples existing in nature and empirical experimentation [13, 38]. Fermentation occurs naturally, readily and quite promptly in tropical regions, and these may have been the places where it originated. Fermented foods and drinks available in present times and places are too numerous to list. Wine, cheese, beer, bread and pickles are a few common examples. Others from traditional societies with which western people are familiar include yogurt, *tofu* (soybean curd), sauerkraut, *kimchi* (hot pickled cabbage, a Korean specialty), tea, and various liquid sauces used as condiments that are made by Asians from soybeans and fish.

Most foods that are fermented preserve for later consumption a fresh item that would otherwise be spoiled or wasted if it comes into season in quantities too great to be eaten by the local population at the height of its availability. Fermentation transforms it into a product which will keep for much longer periods. Many of these foods are protein sources. Cheese is an obvious instance. A few less well-known examples illustrate the nutritive value of these products, prepared by traditional, indigenous methods as cottage industries [4, 21]. Many of them are strongly flavored and odored, attributes that are perhaps desirable among traditional peoples, since these foods are often taken as condiments or relishes in small amounts to accompany large servings of bland staples such as cereals, roots, or plantains.

In Malaysia, Indonesia, Thailand and the Philippines a fermented food is made from small shrimp or prawns. Mixed with salt, these whole animals are left in piles outdoors for several days to 'work' in the ambient heat, being pounded intermittently into a paste in a mortar [124]. Adventitiously introduced bacteria appear to hasten the fermentation process. Variously called *belacan, trasi, chinchalok,* or *bagoong,* this strong-smelling protein product, to which spices may be added, is an ingredient in cooking curries and vegetables, or combined with chile peppers and other spices as a side relish [124]. Since the carapace (outer shell) of the animal is included, the paste, taken nearly every day, contributes considerable to the calcium intakes of these usual nonmilk drinkers.

East Coast Malays (in Malaysia) prepare a similar relish from small whole fish such as anchovies (*Stolephorus* species) which is called *budu.* The fish are salted and left to ferment for a month in a large covered jar, in a manner somewhat like the production of the Vietnamese fish sauce called *nuoc mam.* However, instead of drawing off the resulting liquid from the bottom of the vessel as is done in Vietnam, Malays take the entire fermented mass to be used as part of a side dish or relish, or to eat as a sauce with rice and other foods at meals. The bones of the fish contribute calcium to the mixture, and other minerals and vitamins present in the fish are also available, along with good quality animal protein [*Wilson,* unpublished data].

Since yogurt is one fermented food that does not keep for long periods, people in the Middle East have developed several 'compound' foods utilizing yogurt as one ingredient [8, 18]. Yogurt cheese *(lebneh)* is mixed with parboiled wheat *(burgul).* (The making of burgul, also known as *bulgur,* with other variant spellings, by parboiling or otherwise cooking wheat is another indigenous cultural preparation practice that enhances nutritional quality of a food, in this case driving the minerals and vitamins from the wheat germ into the starchy endosperm to become available to the consumer after the milling is completed.) The mixture, called *kishk,* is left to ferment at room temperature for several days before being shaped into balls and sundried [8]. (Other yogurt products with good keeping qualities are yogurt butter (source of a shortening, *semneh*) and yogurt buttermilk, made by churning yogurt.) Interest in the nutritive value of this milk-cereal mix has led Mexican authorities to investigate transfer of the technology to that country so that the food could be used to alleviate protein-energy malnutrition [18].

A number of other legumes besides the soybean are fermented by traditional peoples. *Gupta* et al. [47] reported that low income people living under semidesert conditions in the State of Haryana in India maintained good calorie and protein intake by pickling or drying the unripe pods of a leguminous tree *(Prosopis cineraria)* locally known as *jant.* The nutrient composition compared favorably with that of cowpeas. In Ghana and Nigeria the bean of another leguminous tree, *Parkia biglobosa,* is fermented to make a high-protein supplement that has traditionally been added to soups or stews [4, 21]. After being boiled for up to 24 h to soften the seedcoat, the beans are placed in leaf-lined shallow holes or baskets and left to ferment for 3 days with the aid of environmentally present bacteria. Like the Southeast Asian prawn paste described above, the mass is sundried before being pounded into a paste and formed into balls or other shapes and dried further. The product, called *dawadawa* or *dadawa* by the Hausa, *kinda* in Sierra Leone, *netetou* in Gambia, and *soumbala* in French West Africa, will keep for a year in the climate without refrigeration. Also like the Asian prawn paste, it is eaten in few-gram quantities (range 1–17) with a frequency that varies from about 10% of meals in northern Ghana to 90% of meals in northern Togo. Nutrient analysis [21] showed the dry product to be 18–47% protein, 31–43% fat (of which 60% is unsaturated fatty acids), 7–51% iron, and 0.27–2.9% calcium. Its content of lysine, an amino acid limiting in diets in this savannah region of Africa, compares favorably to that of whole egg. Fermentation also increases its riboflavin content, another nutrient often deficient in West Africa, to 0.8 mg/100 ml on a dry weight basis. This good food supplement is also one of the cheapest protein foods of the region.

Sprouting
Closely related to fermentation in its effect on the treated food is sprouting. Both legumes and grains have been used by numerous population groups to make sprouted foods. Most western peoples are now familiar with the varieties of sprouts that result from placing dry beans or seeds in a

warm, moist location for a few days. Alfalfa sprouts have recently become fairly common additions to salads and sandwiches. What may not be widely realized is that the process, which was instituted to make pounding the grain into meal easier, enhances the nutrients, chiefly B vitamins, available to the consumer, as *Culwick* [28] pointed out for the maize eaten in East Africa. Enhancement of the sprouts' content of some nutrients is an unplanned but useful concomitant.

Combinations of Foods

Nutrition scientists have long been aware of the amino acid complementarity that can ensue from combining two differing foods of vegetable origin, such as cereals and legumes, to produce an amino acid pattern comparable to that of high-quality animal protein, the legume supplying the amino acid(s) present in insufficient amounts in the cereal, and vice versa [15]. They may not be so cognizant, however, with the fact that these combinations and the relative amounts in which each food is eaten were arrived at empirically, long ago, by various groups of traditional peoples, on the basis of a cultural definition of what goes together to make a meal that includes approximate proportions of each component that should be taken. It is only in about the last two decades that experimental administration of some of these combinations of foods to laboratory animals has shown how close to needs the traditional combinations and amounts have come. Persons whose diets are based on such combinations have strong feelings about the appropriateness of the components for certain meals, and about the combinations themselves, to the extent that, if one component is missing, they may be indifferent to the remainder, preferring to eat something else instead [30], although they may profess hunger in the absence of these foods [124]. Such attitudes may represent sound nutritional behavior, providing the substitute is nutritionally comparable to that of the temporarily discarded meal in its complete state.

These feelings and the combinations are culturally sanctioned by the group to which the individuals belong. The 'rightness' of the diet is not questioned. The numbers of combinations are many and they are found worldwide. Most involve a cereal and a leguminous seed, but a few utilize small amounts of an animal protein source such as a milk product or fish to complement and complete the grain-based meal. The Latin American maize corn and beans combination is well known. People of European origins may not realize that wheat bread and cheese is a comparable one for them. Macaroni (or other pasta) and cheese, and pizza and lasagna made without meat, are sophisticated refinements of this pattern. Rice and beans, rice and peas, rice or wheat and lentils are others. The wheat and yogurt combinations cited earlier [8] are also examples of complementarity, as is the rice and fish cuisine of East and Southeast Asia [24]. The fish, like the legumes, supply the lysine that is insufficient in the grain to meet human needs, although the small, anchovy-like animals studied in Southeast Asia [24] are usually taken in much lesser quantities relative to the amount of rice consumed than are most legumes. The groundnut (peanut, *Arachis hypogoea*), which is eaten as a spiced sauce with a cereal staple in regions as far apart as West Africa and Indonesia, provides lysine to the consumer in the same manner as the other pulses cited, and the cereal supplies the sulfur-containing amino acids (methionine and cystine) that are inadequate in all these leguminous foods.

These combinations of cereal plus legume or cereal plus small amounts of animal food have served in many times and places as the base of the cuisine, i.e. the cereal-plus-protein meal is what *Bennett* [10] and subsequently *Jerome* [59] have termed the core of the diet. These two items, a starch staple and a protein source, whether vegetable or animal, are usually the basis of the main meal that may be called dinner, or its equivalent, and it will be what its consumers will always think of when asked if they have eaten. Other items may be taken along with these core foods, but the latter are always present. If this base is disrupted, the resulting altered meal may be considered snacking, or those experiencing substitutions for usual food items may aver they have not eaten, or say they feel hunger, even though they may have taken a physiologically and nutritionally satisfactory substitute for the missing foods. Thus, through cultural definitions nutritionally sound food practices have arisen.

Beliefs and Practices Surrounding Foods

Foods for Vulnerable Groups
Despite the introductory statement that suggests there are few positive beliefs or practices that deal with foods or their properties, some do exist. As with prohibitions or taboos against foods, many of the more positive ideas also devolve upon the vulnerable population segments, reproducing women and their nursing or toddler children. A few societies feed colostrum to their newborns [74, 104, 121]. Recent research indicates that the practice

may be more widespread than has hitherto been believed [*Raphael,* unpublished data]. Many population groups recognize the value of breast milk to the infant [7, 74, 86]. Although reports are plentiful of weaning to foods that are inadequate quantitatively or qualitatively, *Jelliffe* et al. [57] found that the Hadza hunter-gatherers of Tanzania fed babies marrow, soft (animal) fat and a drink made from the seed of the baobab tree *Adansonia digitata),* which is an excellent source of ascorbic acid. *Mackenzie* [73], studying the Polynesian Cook Islanders, noted that these people, who value a fat baby, have 'nutritionally sound' foods suitable for small children, who are fed neither least nor last. *Fry* [41], working with the same population nearly 20 years earlier, found that baby foods on these islands were soft mixtures that included coconut, taro, mashed papaya and banana, and premasticated foods. They were offered at 9–10 months of age, about the end of the breast-feeding period (see also *Ferro-Luzzi* [36] on former infant diets in the Society Islands). Eskimos, another group that nursed children for long periods, up to 2–3 years, also gave children premasticated adult foods before and after weaning [103], until public health authorities discouraged the practice to control spread of infectious diseases. American Indians had the same practice, and the same experience of its interruption [93]. *Stapleton* [110] found the Igorot people of Luzon, the Philippines, also supplemented infant breast-feeding with prechewed adult food.

The almost worshipful regard in which traditional peoples have held their staple often leads to its early introduction and overfeeding to the young child, a complaint that *Jelliffe and Bennett* [55] made about plantain among the Bantu Baganda. This writer has noted the same behavior with respect to cooked rice for Malay infants [124]. Modifications of the staple or cuisine, however, to suit the less developed digestive system of the baby may not be harmful, and may even be beneficial. Mexican-American mothers used to give to little children too young to eat usual table food bits of maize corn tortilla soaked in the water in which the beans were cooked [*Wilson,* unpublished data]. In this instance, early introduction of the child to the basic components of the cuisine is of some if not substantial nutritional worth. *De Garine* [42] has suggested that the reason traditional people introduce the staple to the young child very early is because it is considered safe for consumption by everyone, being equated with the human body, unlike 'strong' adult foods, which are considered capable of causing a shock to the infant if they are introduced too swiftly. This, he thinks, is also why white, bland foods, resembling human milk, are acceptable for infant feeding.

Prolonged breast-feeding (up to as much as several years of life of the child), still practiced by some traditional societies, when it is supplemented by other foods assures the child of a safe, uncontaminated source of animal protein. *Slome* [107] has pointed out that the long (in western eyes) delay in weaning children among 'primitive' peoples meant the severage from the breast and sole reliance on other foods occurred when the child was both more emotionally and physiologically developed. Taboos against sexual intercourse during suckling have been seen by one nutritionist as a protection against the need for weaning the child outside in order to safeguard the growing one within [17]. The culturally stated reasons are of course usually at variance with this explanation. *Di Domenico and Asuni* [31] reported a belief from Nigeria that sperm would contaminate the breast milk of a lactating woman, and make it sour, so intercourse was forbidden to the parents in the first year of the infant's life. A cross-cultural survey of breast-feeding among traditional peoples [86] demonstrated that weaning used to be a gradual process, and all cultures following this practice seemed to have an awareness of the value of breast-feeding to the child. *Sanjur* et al. [102], studying feeding of infants in rural Mexico, found that this population segment was treated by a different set of cultural rules with regard to food than the rest of the family, being gradually incorporated into the family and the family diet.

As is the case with the weanling, diets of pregnant women are rather widely thought to be inadequate due to cultural taboos, particularly on animal protein foods, held by many societies to be too 'strong' for someone in such delicate condition. (These foods are most commonly feared to be potential abortifacients.) Some exceptions to this situation exist, however. In Nepal, *dahl* (lentils, *Lens culinaris*), meat, rice and fruits were favored in pregnancy [16]. The Gilbert Islanders of the South Pacific recommended fish and animal foods in pregnancy, deeming them also advisable for lactation [54]. *Millis* [78] found in Malaya, as did *Wilson* [125] more than a decade later, that pregnant women tended to eat a usual, nonpregnant diet, save for some possible magical avoidances (precautions taken in an eclipse, avoiding certain behaviors that might 'mark' the unborn child). Some Malay women indicated they avoided 'sour' foods [125]. Similar attitudes, including magical precautions against eclipses, were found among Mexican-American women [*Wilson,* unpublished data].

In a study of the ecology of malnutrition among the Ngoni tribe of southwestern Tanzania, *Robson* [99] learned that these people reserved millet *(Eleusine coracana)* made into a lightly fermented beer, *togwa,* for pregnant women. It was found [95] to be a valuable source of calcium as well as energy. The Chinese traditionally prescribed powdered antler, another source of calcium, for pregnant and lactating women [84]. The Chinese practice of cooking bones, together with meat, in slightly acid liquid (see above, p. 70) also would be beneficial to pregnant women, with their greater calcium needs. In Senegal, while the pregnant woman was obliged to avoid spicy foods and condiments, she was encouraged to take curdled milk, palm oil, meat and butter with the millet porridge staple. Her husband was willing to pay to satisfy her cravings for specific foods. If they were not satisfied, it was feared the baby would be born with undesirable birthmarks.

The belief that cravings of pregnancy, if unmet, would result in marks on the newborn resembling the food not eaten is rather widespread. *Newman* [85] reported numerous instances in women of several ethnicities and socioeconomic classes attending clinics and private practitioners in northern California. In ancient Polynesia, strongly flavored foods were tabooed to pregnant women, but their cravings were fostered [58]. An Egyptian woman in the second trimester of pregnancy is allowed free choice of foods, including those she craves, because eating is thought to prepare her well for the ordeal of pregnancy and delivery [1]. A number of the foods pregnant women are reported to crave are nutritionally sound and beneficial. Research on cravings and aversions among western women suggests a preference at this time for milk, ice cream and fruits [29, 51]. *Ammar* [1] noted a craving among pregnant Egyptian women for fruits. The British and US researchers [29, 51] suggest that cravings of pregnancy may be related to physiologic alterations in taste and smell sensitivity. *Dickens and Trethowan* [29] also found a desire in these women for 'chewy' substances. Pregnant Malay women, who did not mention cravings in interviews, chewed raw rice [125]. They were not seen to do so at other times. The nutrients in the rice are nutritionally available to the chewer [*Lim*, personal communication].

Although there may be a physiologic basis for the cravings of pregnancy, they are also culturally sanctioned. *Obeyesekere* [87] reported that in Sri Lanka (Ceylon), many foods are culturally defined to be craved. In this Hindu society, their denial might prevent the child's rebirth, and the normally superior husband must obtain the desired foods for his wife, as must the Egyptian father and his neighbors. Not to do so would bring social disparagement to the Egyptian from his wife's family [1]. It would seem that cultural fostering of cravings may function to offset some of the food taboos otherwise visited upon pregnant women. Positive attitudes toward the cravings encourage consumption of the items desired, and are a legitimate social way to get around some of the taboos.

Pica, the eating of dirt or clay, a practice considered by some workers to be a form of craving, has been suggested by several [52, 82, 109] to have cultural origins. Whether pregnant women seek these earths to meet a 'felt' need for minerals is moot, since other, nonnutritive items such as ashes, matches and laundry starch are also reported to be consumed by pregnant women [67, 116]. Clay eating may cause deficiency of some minerals by complexing and binding them [48].

Postpartum food restrictions observed for cultural reasons are seemingly even more widespread than those for pregnancy. Nevertheless, a few societies have treated new mothers rather well. Among the Abasamia, a tribe of western Kenya, new mothers were given fish or meat [88]. In Indonesia, despite taboos for a month after childbirth against fruits, vegetables and many rich foods, women took a soup of the leaves of papaya, spinach and sweet shoots *(Sauropus)* to stimulate milk production [108]. In Egypt, a new mother was given chicken, chicken broth, clarified butter, molasses and other rich dishes [1]. Traditionally, Chinese women in many parts of the globe still observe a postpartum confinement called 'doing the month'. This behavior, ritually prescribed for 30 days, entails restrictions against activity, against exposure to 'cold', and eating 'cold' foods, notably fruits and vegetables. However, 'hot' foods are not only indicated but urged, to the point of overstuffing. These include chicken, ginger, broths and wine [94]. A chicken a day is the usual expected consumption pattern of these women. *Chung* [26] reported that southern Chinese favor giving new mothers pigs' feet cooked in vinegar and sugar, whereas northern Chinese make millet soup for them.

Some galactogogues prescribed for nursing mothers also make useful nutrient contributions. *Kelly* [64] reported that mothers in North Mexico used beer and *atole* (a cereal-based drink) to stimulate lactation. The most common atoles were made of maize meal, chickpeas and oatmeal, and sesame seed was sometimes also incorporated into the atole given pregnant women. Either combination provided good quality vegetable protein to a population that did not routinely consume animal protein sources on a daily basis. In two Indian villages *Misra* [79] found that certain greens, other vegetables, garlic and coconuts were all used as stimulants to lactation. Mothers of the Igorot tribe of the Philippines were urged to eat a variety of foods to increase milk production [110].

The fairly well known taboo against a lactating woman taking spicy foods due to a belief that they would reach the child through her milk and cause him harm has been shown to have some basis in fact. Research has shown that many substances a women ingests reach her breast in 1–2 days after their consumption in food [92]. How much of individual chemicals such as the active principle in chile peppers gets into the milk is yet to be ascertained.

The nutritionally critical postweaning, preschool years are recognized as times for which many societies do not provide or recommend specific foods or feeding patterns. Toddlers and other children able to walk without help may leave the mother's side for long periods and eat a variety of foods that most survey methods do not elicit [122]. This nibbling may be of high carbohydrate foods, or it may be fruits and other wild foods. Relatively little research has thus far been done to determine the away-from-home intake of the younger members of this age group.

Better known are the food practices of somewhat older children. Preadolescent boys in a number of societies are allowed or even encouraged to forage for food. (*Farb and Armelagos* [35] have pointed out that children in western cultures also seek food at nonmeal times, but this activity is usually confined to helping themselves from cupboards and refrigerators, or begging for snacks from a parent or caretaker.) There are sex differences in this food-seeking behavior, females being more likely to stay closer to the adult female food preparer and get bits of food from her, while wandering boys are allowed to gather a variety of edible times [34, 42]. These may include protein foods not eaten by adult males for cultural reasons. *Ross* [101] cites crustaceans and mollusks for Melanesian children. In other cultures they may be insects, frogs, or small rodents.

In many places these foods include wild plants, especially greens. Some 40 years ago *Ashton* [5], studying the diet of the Sotho, a Bantu group of southern Africa, found that wild greens were widely eaten, although only by women and children. *Walker* [117] who earlier noted with regret the lowering of intake of these foods among Africans upon urbanization, with his co-workers has more recently demonstrated [118] that these wild plants tend to be lower in oxalates than commercial substitutes, permitting the consumer greater calcium absorption and retention. Working more recently among the agropastoralist Tswana of the eastern Kalahari region, *Grivetti* [45] found that their children were beginning to lose their ability to recognize wild foods that were edible, although *Fleuret* [37], farther north in Tanzania, noted that wild greens were still essential to meals at all seasons.

Indians in Mexico and Central America used to consume wild greens as parts of a meal [4]. *Messer* [77], who has studied their consumption in Oaxaca, Mexico, notes that they are now culturally defined as 'inedible' for everyone but small boys, who will 'eat everything'.

The cultural definition of fruits as snacks is rather widespread. A corollary to this definition is that they are particularly suitable for children. The two populations studied by this writer (Malays and Mexican Americans) seldom served fruits at meals. Instead, they were considered something to be eaten socially at other times, or while out walking in country areas, where the bearing trees and shrubs would be encountered. These and other foods taken similarly in a variety of regions such as chickpeas and peanuts, foods to 'keep the mouth busy' [76], thus make nutritional contributions not always sought or recorded by researchers.

Finally, a comment of *Grivetti* [45] regarding the Tswana and their food taboos is worth repeating: Food prescriptions or proscriptions that relate to sex, age and social development reduce competition for dietary resources, assure their equal distribution, and sustain their nutritional quality. Similar observations have been made by other workers with regard to taboos on relatively scarce animal protein foods for women and small children that favor adult males [113].

Foods for Illness

It is rather widely held that food restrictions commonly applied to ill persons are detrimental, particularly those for the young child [39, 44]. However, digestive function may well be lowered in illness. Most of us have experienced a temporary desire for withdrawal from certain foods (such as 'cold' fruits or vegetables) during a bout of intestinal influenza or a stomach upset [126]. Some of the dense starch staples basic to many diets may indeed be too 'heavy' for a sick child, particularly if the usual accompaniment is a spicy sauce [42]. The pulp of the baobab fruit, given to such a child in Senegal, is according to *Watson* [119] also a rich source of ascorbic acid, although, as indicated above, the cultural reason it is given, is for its resemblance to (white) mother's milk [42].

'Tea', a European once commented, 'is a drink in Europe, a medicine in the United States.' The soothing effects of tea in digestive disturbances are well known. The tannins it contains have beneficial astringent properties (but see above, p. 73). Hot liquids in general help keep nasal passages moist and thus 'wash out' the viruses causing upper respiratory infections. For this reason, chicken soup is indeed a good recommendation for the temporarily indisposed, and if it includes some of the chicken meat as well, this becomes a reasonably nutritious temporary food.[1] Malays recommend soft boiled rice for upset stomachs (and almost all other ills [124]). This simple carbohydrate food is easily digested, providing a ready source of energy, so it too is an acceptable food for a brief time during indisposition. The nutritional problems that arise from use of these substitute foods are due to their being administered for longer periods rather than as interim emergency measures, for they are generally short in protein and other nutrients, if not in energy.

A few cultures may provide better than normal diets for the ill. Among a rural Philippine population that, like the Malays, attributed health-giving properties to the basic staple rice, efforts were made to add milk, eggs and meat to the diets of those who were sick [115]. *Chung* [26] reported that (mainland) Chinese policy gave preferential treatment to pregnant women and children when eggs, chickens and milk were scarce on the market. In the Andean region of Venezuela people believe that unsatisfied food cravings in any member of the population can result in a nonmagical 'hunger' disease [111]. It is treated by 'nourishing the body' with soda crackers, egg broth, black beer and fruit juice, none of these being harmful nutritionally, and probably in the main beneficial.

Some folk beliefs about foods and health are beginning to be shown to have at least partial bases in facts. Garlic has long been held to have almost magical properties in warding off a range of illnesses, whether it is eaten, or worn on a string round the neck. Only recently have scientific studies been made of its chemical properties. Two different groups of researchers [22, 23] have shown that compounds in the volatile oil of this bulb that are associated with its characteristic odor lower blood levels of glucose and of lipids. Onions, botanically of the same family, have the same oils with the same properties, although to a lesser extent.

A researcher at the University of California has found that eating an apple into which some iron nails have been previously inverted for approximately 24 h provides iron to the apple eater, an old folk remedy given to children thought to be anemic. One apple treated with several nails can

---

[1] *Farb and Armelagos* [35] cite recent research suggesting support for the belief that the sufferer should 'starve a fever', a still-held, modern folk medical tenet. Fever decreases the blood level of iron, which disease organisms need to survive. Decreased intake of food, and thus iron, might indirectly decrease the infection, according to this hypothesis, by lowering the number of viable causative microorganisms. It is not recommended to put this theory to the test.

provide up to 15 mg of iron [*Rosanoff,* personal communication]. (Readers are not recommended to repeat the experiment without guidance, for galvanized, coated, or part-copper nails might produce a toxic apple.) Other research indicates that pectin, found in apple skins, lowers blood cholesterol levels to a greater extent than many other fibers. Thus, an apple a day may help keep the cardiac specialists at bay.

Other Cultural Practices

Some other cultural behaviors are also nutritionally beneficial. Several researchers have noted that the social practice of sharing food widely spreads available nutrients to a large number, while ensuring consumption (particularly of fresh or uncooked foods) while they are at a rather high level of nutrient content [127]. Other aspects of sharing include feasting, which also distributes nutrients widely if in small amounts or irregularly [61]. Some anthropologists have viewed feasting as a means for certain populations to obtain good quality animal protein more or less in amounts needed as supplements, particularly when the usual diet is predominantly of vegetable foods. For many groups meat, especially that of large animals, is reserved for such occasions. Thus, *Rappaport* [97] has seen the periodic pig killing feasts of New Guinea highland tribes as a means of protecting the environment while redistributing pork surplus at a time when the population was in need of high-quality protein. *Koo* [66] reported that the Taiwanese were faithful in observing religious holidays, which occur on average twice a month, because they were good opportunities to eat meat and other high quality foods. *Chun* [25] has argued that the regular, repetitive ceremonies required of village Koreans to honor ancestors, taken together with other life cycles, religions and annual reasons for feasting, provide these villagers with enough animal protein to meet most of their needs for this nutrient, since the principal foods for these occasions are good quality marine animals, and the members of the kin and social networks prescribed by custom to share in each ceremony are wide, overlapping, and interlocking. Some of *Rappaport's* assertions have since been refuted [71], or called in question. Further examination of nutritional contributions of feasts in other settings might be more confirmatory of the other hypotheses.

The Northeast Coast Bantu established a reciprocal network of sharing at feasts and by means of trading that made sure fishermen got garden produce and farmers got fish [43]. Another form of food sharing was demonstrated in East Africa, in former times, in which the 'big people' (chiefs and their wives) ate first, but with restraint, so that there would be enough food for everyone [98]. Food sharing may be an innate social activity. *Dyson-Hudson and van Dusen* [33] found it to occur almost spontaneously in preschool children in day school. *Kavanagh* [63] noted it among monkeys.

Snacking, often done in a social context, i.e. in informal gatherings with others, might be considered the socialization of a physiologic desire to experience repetitive oral activities [91]. Traditional snacks of legumes and fruit are beneficial additions to the diets of those who eat them, although the items consumed are not infrequently not considered to be 'food', but components of a pasttime (see above, p. 77). Snacking may offset some inequities in food distribution at meals among family members [123].

Use of wild plants as parts of dietaries gave their consumers a rather wide resource base, in addition to good nutrient sources, as suggested earlier [19, 100]. Sometimes these 'free' foods were held in good esteem. This seems to have occurred with the soup made in the Caribbean of green leaves and other vegetables, called *callaloo*. To the Black West Indians who consume it, it is a recollection of their African heritage, and thus valued.

Conservatism in resource utilization that has characterized many hunting-gathering and traditional subsistence peoples has indirect nutritional benefits as well [9]. *Orraca-Tetteh* [90] has reported that the Ga people of Ghana forbade taking breem, a ritually esteemed fish, for several months each year, a possible ecologic control on overfishing. Similarly, *Thompson* [112] observed that Fijians used to employ a crop custodian who tabooed fields until the foods were ripe. *Townsend* [113], studying a New Guinea group dependent upon sago, found that they could raise more of their staple than they did, and feed a larger population, but preferred to practice population control to prevent reduction of more limited available wild animal food. The Bayano Cuna Indians of Panama studied by *Bennett* [9] practiced a deliberate cultural isolation from other groups, which helped maintain availability of wild animals used for food. The Baegu forest farmers of the island of Malaita in the South Pacific produced more than enough for their needs, ate well and healthily, and 'conserved their land and resources' [101].

Wide exploitation of available resources is another nutritionally beneficial cultural practice. Although perhaps no group has consumed all the potentially edible items in its environment, a few have come rather close. A classic example is the Otomi Indians of the Mezquital Valley of Mexico, studied more than 30 years ago by a group of food scientists from the Massachusetts Institute of Technology [3], who used '.. every conceivable plant available' to obtain a fairly good diet. *Berlin and Markell* [11] found the Aguaruno Jivaro Indians of Amazonas, Peru, by combining hunting, gathering and fishing with cultivating, obtained a diversified diet of high quality. In the Andes, the diet of the Quechua is qualitatively different from and better than that of Indians living at lower altitudes, due in part to exploitation of a greater ecologic diversity [75]. *Robson and Wadsworth* [100] have pointed out that the diets of 'primitive' aboriginals were extremely varied, while *Draper* [32] has noted that Eskimos in earlier times ate a balanced diet from what was present because there was little else to eat. Their utilization of the animals which were their staples included consumption of the intestinal contents, in which bacteria present had synthesized various vitamins.

Sunworshipping was widespread among prehistoric peoples [40]. The festive celebrations at the time of the spring and autumnal solstices and the midwinter fire rituals still practiced among northern Europeans (burning the Christmas Yule log, for example) are remnants of ancient attitudes of gratitude to the sun and its annual reappearance or rebirth. Some researchers have suggested that sunworship may in part have been early man's recognition of the healthful benefits on bone and teeth of what is now known as vitamin D, manufactured on the surface of the skin from a derivative of cholesterol [15, p. 139].

Biocultural Factors

Scientific research is still uncovering or elucidating mechanisms of considerable complexity with components affecting the human nutritional state beneficially that have been recognized empirically and taken advantage of by various population and ethnic groups. One example already cited (see above, pp. 76–77) is the widespread dietary combination of cereal grains and legumes or grain products and small amounts of animal protein food such as fish or cheese that, taken together in customary amounts, provide better quality protein than either item could alone, through the additive values of the constituent amino acids. It is probable that, as more traditional diets and their preparation are examined and reported upon, other worthwhile cultural practices that enhance the nutritive value of diets will become known.

The intricate biocultural process of breast-feeding is another case in point, in which a complex feedback system exists among the mother's hormone system and the baby's suckling [96]. Vigorous sucking on the part of the infant, which appears initially to be a reflexive reaction but later becomes 'learned', stimulates release of prolactin by the mother's pituitary gland to activate milk formation in the mother's breast. The sucking also stimulates release of another pituitary hormone, oxytocin, which initiates the 'let-down' impulse that causes the secreted milk to pass into the lacteals and out through the nipple. Regular and continued sucking by the baby ensures a regular, continued supply of milk. Lactation inhibits estrogen production by the ovaries, probably because prolactin prevents the hypothalamus from forming the luteinizing hormone that stimulates the ovaries to release eggs. The effect is not complete or total. However, mothers who breast-feed solely on demand, for the first several months of the infant's life are less likely to conceive than those who suckle less often, or who introduce supplementary foods while still nursing. *Jelliffe and Jelliffe* [56] have pointed out that the rapid 'bonding' between the nursing mother and suckling baby, due to their interdependence in bringing about the milk flow, increased the child's chances for survival in societies where breast milk is the only food available for a newborn.

The composition of human milk is 'programmed' to fit the nutritional needs of the rapidly growing newborn. It is, for example, calorically dense, 50% of its energy being in the form of readily digested fat. This means that the baby gets the energy it needs in an amount of fluid that its small digestive system can accommodate. The quality of the fat in mother's milk is correct for production of prostaglandins and for development of the intestinal tract. The anti-infective properties of human milk such as IgA are higher than those of other animals. It contains lactoferrin (transferrin) in amounts that explain why breast-fed infants do not become iron-deficient. A similar compound has been found for zinc. Both compounds stick to the brush border of the intestinal tract [*Barness,* personal communication], apparently thus aiding absorption of these elements. This finding may explain the observation that infants absorb efficiently the rather small amount of iron in breast milk [72].

Human milk contains other nutrients in amounts the newborn needs. It provides, for example, well above the Recommended Dietary Allowance for ascorbic acid when taken in usual amounts. It is rich in vitamin E, and contains other nutrients in amounts the new baby has been shown to need.

One other health benefit to breast-feeding is the position in which the child is held. Reclining helps drain the Eustachian tubes, to alleviate or prevent ear problems such as otitis [*Barness,* personal communication].

Recent studies [46, 49] have shown that the composition of the milk of mothers of premature infants is not the same as milk of mothers of babies born at term, and suggest that it is most suited to the altered needs of children born before the usual time, for a number of key nutrients. The nitrogen of preterm human milk is sufficiently higher than milk of mothers of term babies to provide the prematures with two thirds more protein than would term milk. Sodium and chloride are also significantly higher in preterm milk, as magnesium. Iron tends to be higher. These workers [69] noted that the composition of this preterm milk resembles that of colostrum, and that, if fed to a preterm child, the nutrients would be present in amounts more than sufficient to have maintained growth in the uterus had the infant remained in the womb and been nourished transplacentally.

Although immunoglobulin (IgG) crosses the placenta, and the human fetus can also synthesize this protein by the second half of intrauterine life [80], those cultures that give the child suck in the first few hours or the first day or two of life provide it with a further store of immunity factors from the colostrum, since it is high in globulins with immune properties, particularly IgA, which protects the child against viruses, *Escherichia coli,* and other pathogens [*Barness,* personal communication]. Apparently in the first 2 days of life these proteins are not broken down intestinally but absorbed whole by the infant, another biological adaptation. Like preterm milk, colostrum has more protein than term milk, a lower caloric density and a high level of phospholipids. With its higher vitamin content it is tailored to the unique needs of the neonate in the first days outside the womb.

Biological development affecting nutritional state seems sometimes to have been 'known' without being understood, or the underlying biochemical mechanisms had not yet been explored and explained. Thus, Jewish male infants, who routinely undergo circumcision according to Hebraic law, are not so operated on before the eighth day of life. Research in this century has shown that the newborn, who has only a modest store of the clotting factor vitamin K, begins to manufacture his own supply of this vitamin after the first week postpartum [13]. Some pediatricians would seem to have less folk wisdom than traditional Hebrews, for they prescribe injections of vitamin K for newborns.

A number of genetic adaptations have been postulated to have occurred in response to changes in diets. The best known, perhaps, is the persistence of lactase, the enzyme that digests milk sugar, in persons above the age of weaning [49], particularly those of Northern European ancestry. It has been hypothesized [105] that this change is due to continued milk drinking in dairying groups. The extended existence of ability to utilize lactose is advantageous in climes where vitamin D from sunlight is limited, since lactose also assists intestinal absorption of calcium [49]. *Neel* [83] has suggested that the ease with which certain American Indian and Polynesian populations become diabetic and overweight is a now maladaptive response to regular food intake and decreased activity that was once functional when food supply was irregular and varied greatly in amount, and people performed more intense physical work. *Ho* et al. [50] have postulated a somewhat similar homeostatic ability to adapt to alternating hypo- and hypercholesterolemia among Alaskan Eskimos to accommodate changes in amounts of their often high-fat, meat diet.

Conclusions

This review was undertaken to present some published or otherwise known examples of culturally determined behaviors that have nutritionally beneficial consequences, in order to increase awareness that the rather widely held view that taboos and other cultural beliefs and practices are heavily detrimental to nutritional well-being is not necessarily always so. It is indeed true that many beliefs and practices have disadvantageous consequences for nutritional status of those to whom the restrictions or behaviors are intended or applied [92]. The foregoing discussions are intended to provide a partially counterbalancing view with the hope that, along with the insight that cultural practices are not inevitably or always dietarily bad, there will be wider understanding of the anthropologist's observation that, to their practitioners, culturally determined customs and beliefs have an internal consistency supportive of the society which created and abides by them.

References

1   Ammar, H.: Growing up in an Egyptian village. Silwa, province of Aswan (Routledge & Kegan Paul, London 1954).
2   Anderson, E.N., Jr.; Anderson, M.L.: Modern China: south; in Chang, Food in Chinese culture: anthropological and historical perspectives, pp. 317–382 (Yale University Press, New Haven 1977).

3   Anderson, R.K.; et al.: A study of the nutritional status and food habits of Otomi Indians. Am. J. publ. Hlth *36:* 883–903 (1946).

4   Anonymous: A 'new' protein food from a traditional process. Nutr. Rev. *39:* 9–11 (1980).

5   Ashton, E.H.: A sociological sketch of Sotho diet. Trans. R. Soc. S. Afr. *27:* 147–214 (1939).

6   Baker, P.T.; Mazess, R.B.: Calcium: unusual sources in the highland Peruvian diet. Science *142:* 1466–1467 (1963).

7   Balen, H. Van; Ntabomvura, V.: Methods of birth spacing, maternal lactation and postpartum abstinence in relation to traditional African culture. J. trop. Pediat. Environ. Child Hlth *22:* 50–52 (1976).

8   Basson, P.: Women and traditional food technologies: changes in rural Jordan. Ecol. Food Nutr. *11:* 17–23 (1981).

9   Bennett, C.F.: The Bayano Cuna Indians, Panama: an ecological study of livelihood and diet. Ass. Am. Geog. Ann. *52:* 32–50 (1962).

10 Bennett, J.W.: Food and culture in southern Illinois. Am. Sociol. Rev. *7:* 645–660 (1942).

11 Berlin, E.A.; Markell, E.K.: An assessment of the nutritional and health status of an Aguaruna Jivaro community, Amazonas, Peru. Ecol. Food Nutr. *6:* 69–81 (1977).

12 Bradfield, R.B.; Lauriault, J.: Diet and food beliefs of Peruvian jungle tribes. 1. The Shipibo (monkey people). J. Am. diet. Ass. *39:* 126–128 (1961).

13 Braidwood, R.J.: Symposium: did man once live by beer alone? Am. Anthrop. *55:* 515–526 (1953).

14 Brewer, T.J.: Food practices of some Samoans in Los Angeles County (Department of Health Services, County of Los Angeles, Los Angeles 1973).

15 Briggs, G.M.; Calloway, D.H.: Nutrition and physical fitness; 10th ed. (Saunders, Philadelphia 1969).

16 Brown, M.; Worth, R.M.; Shah, N.K.: Food habits and food intake in Nepal. Trop. Geog. Med. *20:* 217–224 (1958).

17 Burgess, A.: Traditional systems of child care. Some implications for the child health workers in the West Pacific region. Hlth Educ. J. *15:* 99–108 (1957).

18 Cadena, M.A.; Robinson, R.K.: The acceptability of yoghurt-cereal mixtures to a rural community in Mexico. Ecol. Food Nutr. *8:* 169–174 (1979).

19 Caldwell, M.J.: Ascorbic acid content of Malaysian leaf vegetables. Ecol. Food Nutr. *1:* 313–317 (1972).

20 Calloway, D.H.; Giauque, R.D.; Costa, F.M.: The superior mineral content of some American Indian foods in comparison to federally donated counterparts. Ecol. Food Nutr. *3:* 203–211 (1974).

21 Campbell-Platt, G.: African locust bean (Parkia species) and its West African fermented food product, dawadawa. Ecol. Food Nutr. *9:* 123–132 (1980).

22 Chang, M.L.W.; Johnson, M.A.: Effect of garlic on carbohydrate metabolism and lipid synthesis in rats. J. Nutr. *110:* 931–936 (1980).

23 Chi, M.S.; Koh, E.T.; Stewart, T.J.: Effects of garlic on lipid metabolism in rats fed cholesterol or lard. J. Nutr. *112:* 241–248 (1982)

24 Chong, Y.H.; Soh, C.C.: The protein nutritive quality of ikan bilis *(Stolephorus* spp.). Med. J. Malaya *22:* 230–233 (1966).

25 Chun, K.-S.: We are well nourished by virtue of our ancesors: ancestor worship and nutrition in a Korean village. Ecol. Food Nutr. (in press).

26 Chung, A.W.: Breast-feeding in a developing country: The Peoples Republic of China; in Raphael, Breast-feeding and food policy in a hungry world, pp. 81–86 (Academic Press, New York 1979).

27 Committee on Technological Innovation, Board on Science and Technology for International Development: underexploited tropical plants with promising economic values, pp. 20–22 (National Academy of Sciences, Wash. 1975).

28 Culwick, G.M.: Nutrition in East Africa. Africa *14:* 401–410 (1944).

29 Dickens, G.; Trethowan, W.H.: Cravings and aversions during pregnancy. J. psychosom. Res. *15:* 259–268 (1971).

30 Dickins, D.: Some effects of a white cornmeal shortage. J. Am. diet. Ass. *21:* 287 (1945).

31 Di Domenico, C.M.; Asuni, J.B.: Breast-feeding practices among urban women in Ibadan, Nigeria; in Raphael, Dana, Breast-feeding and food policy in a hungry world, pp. 51–57 (Academic Press, New York 1979).

32 Draper, H.H.: The aboriginal Eskimo diet in modern perspective. Am. Anthrop. *79:* 309–316 (1977).

33 Dyson-Hudson, R.; Dusen, R. van: Food-sharing among young children. Ecol. Food Nutr. *1:* 319–324 (1972).

34 DuBois, C.: The people of Alor: a socio-psychological study of an East Indian island (Harper & Row, New York 1961).

35 Farb, P.; Armelagosa, G.: Consuming passions. The anthropology of eating (Houghton, Mifflin, Boston 1980).

36 Ferro-Luzzi, G.: Food patterns and nutrition in French Polynesia. Am. J. clin. Nutr. *11:* 299–311 (1962).

37 Fleuret, A.: The role of wild foliage plants in the diet: a case study from Lushoto, Tanzania. Ecol. Food Nutr. *8:* 87–93 (1979).

38 Forni, G.: The origin of grape wine: a problem of historical-ecological anthropology; in Arnott, Gastronomy: the anthropology of food and food habits, pp. 67–78 (Mouton, The Hague 1976).

39 Foster, G.M.: Social anthropology and nutrition of the pre-school child; in Pre-school child malnutrition: primary deterrent to human progress, National Research Council publ. 1282, pp. 258–266 (National Academy of Sciences, Wash. 1966).

40 Frazer, G.J.: The golden bough: a study in magic and religion; abridged ed., reprinted (Macmillan, New York 1963).

41 Fry, P.C.: Dietary survey on Rarotonga, Cook Islands. 3. Feeding practices and growth of Rarotonga children from birth through six years. Am. J. clin. Nutr. *5:* 634–643 (1957).

42 Garine, I. de: The socio-cultural aspects of nutrition. Ecol. Food Nutr. *1:* 143–163 (1971).

43 Gerlach, L.P.: Socio-cultural factors affecting the diet of the Northeast Coast Bantu. J. Am. diet. Ass. *45:* 420–424 (1964).

44 Gonzales, N.L.S. de; Scrimshaw, N.S.: Public health significance of child feeding practices observed in a Guatemalan village. J. trop. Pediat. *3:* 99–104 (1957).

45 Grivetti, L.E.: Nutritional success in a semi-arid land: examination of Tswana agro-pastoralists of the eastern Kalahari, Botswana. Am. J. clin. Nutr. *31:* 1204–1220 (1975).

46   Gross, S.J.; David, R.J.; Bauman, L.; Tomarelli, R.M.: Nutritional composition of milk produced by mothers delivering preterm. J. Pediat. *96:* 641–644 (1980).
47   Gupta, M.C.; Gandhi, B.M.; Tandon, B.N.: An unconventional legume – *Prosopis cineraria.* Am. J. clin. Nutr. *27:* 1035–1036 (1974).
48   Halsted, J.A.: Geophagia in man: its nature and nutritional effects. Am. J. clin. Nutr. *21:* 1384–1393 (1965).
49   Harrison, G.G.: Primary adult lactase deficiency: a problem in anthropological genetics. Am. Anthrop. *77:* 812–835 (1975).
50   Ho, K.-J.; Mikkelson, B.; Lewis, L.A.; Feldman, S.A.; Taylor, C.B.: Alaskan Arctic Eskimos: responses to a customary high fat diet. Am. J. clin. Nutr. *25:* 737–745 (1972).
51   Hook, E.B.: Dietary cravings and aversions during pregnancy. Am. J. clin. Nutr. *31:* 1355–1362 (1978).
52   Hunter, J.M.: Geophagy in Africa and in the United States: a culture-nutrition hypothesis. Geog. Rev. *63:* 170–195 (1973).
53   Institute of Nutrition of Central America and Panama; Interdepartmental Committee on Nutrition for National Defense: Nutritional Evaluation of the Population of Central America and Panama, 1965–1967. Publ. No. (HSM) 72-8120 (Department of Health, Education and Welfare, Washington 1971).
54   Jansen, A.J.J.: Malnutrition and child feeding practice among the Gilbertese. J. trop. Pediat. Environ. Child Hlth *23:* 161–184 (1977).
55   Jelliffe, D.B.; Bennett, F.J.: Cultural and anthropological factors in infant and maternal nutrition. Fed. Proc. *20:* suppl. 7, pp. 185–187 (1961).
56   Jelliffe, D.B.; Jelliffe, E.F.P.: Human milk in the modern world: psychosocial, nutritional, and economic significance (Oxford University Press, Oxford 1978).
57   Jelliffe, D.B.; Woodburn, J.; Bennett, F.J.; Jelliffe, E.F.P.: The children of the Hadza hunters. J. Pediat. *60:* 908–913 (1962).
58   Jelliffe, E.F.P.; Jelliffe, D.B.: Children in ancient Polynesian Hawai. Clinica. pediat. *3:* 604–613 (1964).
59   Jerome, N.W.: Diet and acculturation. The case of Black-American immigrants; in Jerome, Kandel, Pelto, Nutritional anthropology, pp. 275–325 (Redgrave Publishing, Pleasantville 1980).
60   Jones, W.O.: Manioc in Africa (Stanford University Press, Stanford 1959).
61   Katona-Apte, J.: Dietary aspects of acculturation: meals, feasts, and fasts in a minority community in South Asia; in Arnott, Gastronomy: the anthropology of food and food habits, pp. 315–326 (Mouton, The Hague 1976).
62   Katz, S.M.; Hediger, M.L.; Valleroy, L.A.: Traditional maize processing in the New World. Science *184:* 765–773 (1974).
63   Kavanagh, M.: Food-sharing behavior within a group of Douc monkeys *(Pygathrix nemaeus nemaeus).* Nature, Lond. *239:* 406 (1972).
64   Kelly, I.: Folk practices in North Mexico (University of Texas Press, Austin 1965).
65   Konlande, J.E.; Robson, J.R.K.: The nutritive value of cooked camas as consumed by Flathead Indians. Ecol. Food Nutr. *1:* 193–195 (1972).
66   Koo, L.C.: Traditional Chinese diet and its relationship to health. Kroeber Anthropol. Soc. Papers *47:* 116–147 (1975).
67   Lackey, C.J.: Pica – a nutritional anthropology concern; in Bauwens, The anthropology of health, pp. 121–129 (Mosby, St Louis 1978).

68 Laguna, J.; Carpenter, K.J.: Raw versus processed corn in niacin-deficient diets. J. Nutr. *45:* 21–28 (1951).

69 Lemons, J.A.; Moye, L.; Hall, D.; Simmons, M.: Differences in the composition of preterm and term human milk during early lactation. Pediat. Res. *16:* 113–117 (1982).

70 Lopez de Romana, G.; Graham, G.G.; Mellits, E.D.; Maclean, W.C., Jr.: Utilization of the protein and energy of the white potato by human infants. J. Nutr. *110:* 1849–1857 (1980).

71 McArthur, M.: Pigs for the ancestors: a review article. Oceania *45:* 87–123 (1974).

72 McMillan, J.A.; Landaw, S.A.; Oshi, F.A.: Iron sufficiency in breast-fed infants and the availability of iron from human milk. Pediatrics, Springfield *58:* 686–691 (1974).

73 Mackenzie, M.: Who is a good mother? Ethnomedizin *4:* 7–22 (1976).

74 Mata, L.J.: The children of Santa Maria Cauque: a prospective field study of health and growth (MIT Press, Cambridge 1978).

75 Mazess, R.B.; Baker, P.T.: Diet of Quechua Indians living at high altitude: Nunoa, Peru. Am. J. clin. Nutr. *15:* 341–351 (1964).

76 Mead, M.: Nutrition; in Mead, Cultural patterns and technical change, pp. 194–204 (New American Library, New York 1955).

77 Messer, E.: The ecology of vegetarian diet in a modernizing Mexican community; in Fitzgerald, Nutrition and anthropology in action, pp. 117–124 (van Gorcum, Assen 1977).

78 Millis, J.: Modifications in food selection observed by Malay women during pregnancy and after confinement. Med. J. Malaya *12:* 139–144 (1958).

79 Misra, R.: Breast-feeding and weaning in two Indian villages; in Raphael, Breast-feeding and food policy in a hungry world, pp. 37–44 (Academic Press, New York 1979).

80 Moghissi, K.K.: Relationship of maternal amino acids and proteins to infant growth and mental development; in Mosghissi, Evans, Nutritional impacts on women. Throughout life with emphasis on reproduction, pp. 86–106, see esp. pp. 88–89 (Harper & Row, Hagerstown 1977).

81 Mortimer, W.C.: History of coca: the 'divine plant' of the Incas; reprinted (And/Or Press, San Francisco 1974).

82 Mustacchi, P.: Cesare Bressa (1785–1836) on dirt eating in Louisiana: a critical analysis of his unpublished manuscript De la dissolution scorbutique. J. Am. med. Ass. *218:* 229–232 (1971).

83 Neel, J.V.: Diabetes mellitus: a 'thrifty' genotype rendered detrimental by 'progress'? Am. J. hum. Genet. *14:* 353–362 (1962).

84 Newcombe, K.: Apparent consumption and socio-economic distribution of nutrients in an urban settlement: Hong Kong. Ecol. Food Nutr. *6:* 9–22 (1977).

85 Newman, L.F.: Folklore of pregnancy: wives' tales in Contra Costa County, California. West. Folklore *28:* 112–135 (1969).

86 Niehoff, A.; Meister, N.: The cultural characteristics of breast-feeding: a survey. J. trop. Pediat. Environ. Child Hlth *18:* 16–20 (1972).

87 Obeyesekere, G.: Pregnancy cravings (dola-duka) in relation to social structure and personality in a Sinhalese village. Am. Anthrop. *65:* 323–342 (1963).

88 Ojiambo, J.A.: A background study of the food habits of the Abasamia of Busia

district, western province, Kenya. A preliminary study, 1965–1966. Nutrition *21:* 216–221 (1967).

89 Oliver, D.: A case of change in food habits in Bjougainville, British Solomon Islands. Appl. Anthrop. *1:* 34–36 (1942).

90 Orraca-Tetteh, R.: Cultures in conflict: when does man's cultural heritage no longer act as a protective device? Prog. hum. Nutr. *2:* 80–86 (1978).

91 Oswald, I.; Merrington, J.; Lewis, H.: Cyclical 'on demand' oral intake by adults. Nature, Lond. *225:* 959–960 (1970).

92 Petrakis, N.L.; Mason, M.L.; Doherty, M.; Dupuy, M.E.; Sadee, G.; Wilson, C.S.: Effects of altering diet fat on breast fluids in women. Fed. Proc. *36:* 1163 (1977).

93 Pijoan, M.; Elkin, C.A.: Secondary anemia due to prolonged and exclusive milk feeding among Shoshone Indians. J. Nutr. *27:* 67–75 (1944).

94 Pillsbury, B.L.K.: 'Doing the month': confinement and convalescence of Chinese women after childbirth. Soc. Sci. Med. *12:* 11–22 (1978).

95 Platt, B.S.: Tables of representative values of foods commonly used in tropical countries. Medical Research Council special report series No. 302, p. 6 (HMSO, London 1962).

96 Population Information Program: Breast-feeding, fertility, and family planning. Population reports, series J, No. 24: J527–J575, see esp. pp. J542–J547 (Johns Hopkins University Press, Baltimore 1981).

97 Rappaport, R.A.: Pigs for the ancestors: ritual in the ecology of a New Guinea people (Yale University Press, New Haven 1967).

98 Read, M.: Native standards of living and African culture change. Illustrated by examples from the Ngoni highlands of Nyasaland. Africa *11:* suppl. (1938).

99 Robson, J.R.K.: The ecology of malnutrition in a rural community in Tanzania. Ecol. Food Nutr. *3:* 61–72 (1974).

100 Robson, J.R.K.; Wadsworth, G.R.: The health and nutritional status of primitive populations. Ecol. Food Nutr. *6:* 187–202 (1977).

101 Ross, H.M.: Bush fallow farming, diet and nutrition: a Melanesian example of successful adaptation; in Giles, Friedlander, The measures of man: methodologies in biological anthropology, pp. 505–615 (Peabody Museum Press, Cambridge 1976).

102 Sanjur, D.M. et al.: Infant feeding and weaning practices in a rural pre-industrial setting. Acta pediat. scand., suppl. 200 (1970).

103 Scott, E.M.: Nutrition of Alaskan Eskimos. Nutr. Rev. *14:* 1–3 (1956).

104 Schack, D.: Nutritional processes and personality among the Guarage of Ethiopia. Ethnology *8:* 292–300 (1969).

105 Simoons, F.J.: The determinants of dairying and milk use in the Old World: ecological, physiological, and cultural. Ecol. Food Nutr. *2:* 83–90 (1973).

106 Simoons, F.J.: Northwest Ethiopia. Peoples and economy (University of Wisconsin Press, Madison 1960).

107 Slome, C.: Culture and the problem of human weaning. J. trop. Pediat. *6:* 23–34 (1960).

108 Soedarmo, P.: Geographic peculiarities of nutrition (Indonesia). Proc. 7th Int. Congr. Nutr., vol. 3, p. 327–301 (1966).

109 Solien, N.L.: A cultural explanation of geophagy. Fla. Anthrop. *7:* 1–9 (1954).

110 Stapleton, L.B.: Some patterns of feeding and support among the Igorot of Sagada; in

Raphael, Breast-feeding and food policy in a hungry world, pp. 45–50 (Academic Press, New York 1979).

111 Suarez, M.M.: Etiology, hunger, and folk diseases in the Venezuelan Andes. J. anthrop. Res. *30:* 41–54 (1954).

112 Thompson, L.: The relations of men, animals, and plants in an island community (Fiji). Am. Anthrop. *51:* 253–267 (1949).

113 Townsend, P.K.: New Guinea sago gathers: a study of demography in relation to subsistence. Ecol. Food Nutr. *1:* 19–24 (1971).

114 Townsend P.K.; Lias, S.-C.; Konlande, J.E.: Nutritive contributions of sago ash used as a native salt in Papua New Guina. Ecol. Food Nutr. *2:* 91–97 (1973).

115 Valdecanas, O.: Barrio central: a study of some social and cultural factors in malnutrition. Philippine J. Nutr. *24:* 223–237 (1971).

116 Vermeer, D.E.; Frate, D.A.: Geophagy in a Mississippi County. Ass. Am. Geog. Ann. *65:* 414–424 (1975).

117 Walker, A.R.P.: Health hazards in the urbanization of the African. Am. J. clin. Nutr. *11:* 551–553 (1962).

118 Walker, A.R.P.; Walker, B.F.; Wadvalla, M.: An attempt to measure the availability of calcium in edible leaves commonly consumed by South African Negroes. Ecol. Food Nutr. *4:* 125–130 (1975).

119 Watson, J.D.: Ascorbic acid content of plant foods in Ghana and the effects of cooking and storage on vitamin content. Ecol. Food Nutr. *4:* 207–213 (1976).

120 Werge, R.W.: Potato processing in the central highlands of Peru. Ecol. Food Nutr. *7:* 229–234 (1979).

121 Wickes, I.G.: A history of infant feeding. 1. Primitive people: ancient works and Renaissance writers. 2. Seventeenth and eighteenth centuries. Archs Dis. Child. *28:* 151–158, 232–240 (1953).

122 Wilson, C.S.: Child following: a technic for learning food and nutrient intakes. J. trop. Pediat. Environ. Child Hlth *20:* 9–14 (1974).

123 Wilson, C.S.: Culture-made nutritional inadequacies; in Jelliffe, Jelliffe, pp. 443–460 (Plenum Press, New York 1982).

124 Wilson, C.S.: Food beliefs and practices of Malay fishermen: an ethnographic study of diet on the East Coast of Malaya; PhD diss., Berkeley (1970).

125 Wilson, C.S.: Food taboos of childbirth: the Malay example. Ecol. Food Nutr. *2:* 267–274 (1973).

126 Wilson, C.S.: Proposed: that 'hot' and 'cold' food beliefs have pharmacological bases in fact; in Cunningham, Proc. Postplenary Session on Medical Anthropology, Poona. Xth Int. Congr. of Anthropological and Ethnological Sciences (Cambridge University Press, Cambridge, in press).

127 Woodburn, J.: An introduction to Hadza ecology; in Lee, DeVore, Man the hunter, pp. 49–55 (Aldine/Atherton, Chicago 1968).

C.S. Wilson, MD, Department of Epidemiology and International Health, University of California, San Francisco, CA 94143 (USA)

Wld Rev. Nutr. Diet., vol. 45, pp. 97–132 (Karger, Basel 1985)

Vitamins and Immunocompetence[1]

Richard S. Panush, Jeffrey C. Delafuente[2]

Division of Clinical Immunology, Department of Medicine, College of Medicine,
University of Florida; Veterans Administration Medical Center, Gainesville,
Florida; Department of Clinical Pharmacy, St. Louis College of Pharmacy;
Department of Internal Medicine, St. Louis University School of Medicine,
St. Louis, Mo., USA

Contents

[1] Supported in part by the Florida Chapter of the Arthritis Foundation, the Medical Research Service of the Veterans Administration, and the Roche Research Institute.

[2] The authors appreciated the skilled assistance of *Barbara Gibbs, Alice Cullu,* and *Anne Barry* in helping prepare the manuscript.

Introduction

The immune system, a marvelous and intricate network, differentiates self from non-self and maintains host defenses against foreign substances and pathogens. We have come to appreciate that in addition to complex interrelationships within the immune system, other factors, including pharmacologic agents, hormones, exercise, stress, physiological factors, microorganisms, infectious agents, disease states, foods and food products, nutritional status, and vitamins, may perturb immune responses. We shall review the information available regarding vitamins and the human immune response.

Human Immune System

To facilitate discussion of immunologic effects of vitamins, we shall provide a brief overview of the human immune system, which is composed of two interacting limbs – the humoral immune system and the cellular immune system [*Delafuente* et al., 1981]. Humoral immunity is mediated by an array of proteins, peptides, and small molecules, such as antibodies, complement, kinins, and prostaglandins. Lymphocytes, macrophages, and neutrophils are responsible for cell-mediated immunity (fig. 1).

Humoral Immunity

Immunoglobulins. Immunoglobulins (antibodies) compose a major element of the humoral immune system. Immunoglobulins are protein molecules secreted by B lymphocyte-derived plasma cells. The antigen-binding region of each immunoglobulin molecule is specific for the inducing antigen. The basic antibody molecule comprises two identical large polypeptide chains (heavy chains) and two identical small polypeptide chains (light chains). In humans there are five major classes of immunoglobulins, designated IgG, IgM, IgA, IgD, and IgE. The amino acid sequence in the heavy chains determine the immunoglobulin class.

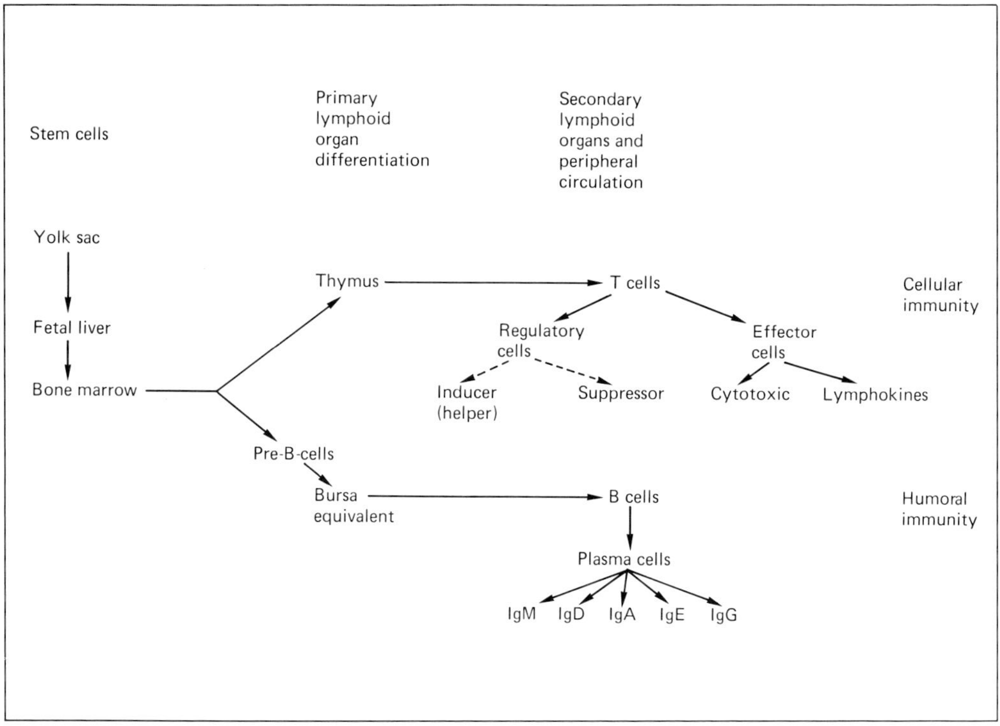

Fig. 1. Pluripotential stem cells differentiate into T or B cells upon being influenced by the thymus or bursa equivalent, respectively. T cells are largely concerned with so-called cellular immunity or delayed-type hypersensitivity, whereas B cells participate in humoral immune processes.

Each of the immunoglobulin classes has differing functions [*Bloch,* 1979]. IgG predominates in serum. Its biologic activities include toxin neutralization, enhancing macrophage and neutrophil phagocytosis, and destruction of bacteria through its ability to activate the complement system. It is also involved in the pathogenesis of many immune complex-mediated diseases. IgM is the first antibody class to appear after a primary exposure to an antigen. Like IgG, IgM can neutralize toxins and activate complement. It also serves as an antigen receptor on B lymphocytes. IgA is found in serum and secretions and offers host protection to areas such as the respiratory tract and gastrointestinal tract. Secretory IgA can prevent bacteria from adhering to mucosal surfaces, thus preventing colonization. IgA can also neutralize toxins. IgD is found in small quantities in serum, and presently its only known role is to serve as an antigen receptor on B lymphocytes. IgE also occurs in very low quantities in serum. IgE can bind to tissue mast cells and blood basophils. When these cells are sensitized with IgE and then come into contact with a specific antigen, the cells release potent mediators that are responsible for allergic manifestations, such as hives, asthma, or anaphylaxis [*Delafuente* et al., 1981].

Complement System. The complement system is a group of at least 21 proteins that interact upon appropriate stimulation to generate biologically active substances, which enhance the inflammatory response and increase defenses against infection. Activation of complement can lead to lysis of hematopoietic cells, bacteria, viruses, and tumors. Several of the complement peptides can enhance phagocytosis, while others increase vascular permeability and augment inflammation [*Delafuente* et al., 1981]. The roles of kinins, prostaglandins, leukotrienes, and products of the clotting and fibrinolytic systems in immune responses are presently undergoing intensive investigation in many laboratories.

Cellular Immunity
Cell-mediated immunity is a specific immune response to antigens mediated by sensitized lymphocytes and macrophages. Many diverse antigens – such as protein, polysaccharide, chemical, bacterial, viral, tumor, and transplantation antigens – may stimulate the immune system [*Delafuente* et al., 1981].

Lymphocytes. The lymphoid system is comprised of the stem cell pool, the primary lymphoid organs (thymus and bursa equivalent), and the secondary lymphoid organs (lymph nodes, spleen, and Peyer's patches). Pluripotent stem cells can differentiate in the primary lymphoid organs to become mature B and T lymphocytes. A schematic representation of lymphocyte development is illustrated in figure 1.

T lymphocytes originate in the thymus, are the major cell type responsible for cell-mediated immunity, and can be defined by unique phenotypes. T lymphocytes play a primary role in immunologic responses such as delayed cutaneous hypersensitivity, resistance to viral and fungal infections, transplantation rejection, and tumor resistance. When T lymphocytes encounter certain antigens, they can become either regulatory or effector cells. Regulatory T cells can be helper cells and augment other T and B cells or can be suppressor cells that can inhibit other T and B cells. Effector T cells can be cytotoxic to other cell types or can produce soluble substances (lymphokines), which can modulate the activities of other cells involved in the immune response.

B lymphocytes are the precursors to plasma cells that are responsible for antibody production. In chickens, B cells differentiate in the bursa of Fabricius, an area of epithelial tissue in the gut. The precise location of human B lymphocyte differentiation is unknown.

T and B lymphocytes can be activated in vitro with specific antigens and nonspecific stimuli. Phytomitogens are commonly used in vitro as nonspecific activators of lymphocytes. Phytomitogens, such as pokeweed mitogen, concanavalin A, and phytohemagglutinin, will cause lymphocytes to transform into blast cells. Radiolabeled thymidine is usually added to lymphocyte cultures to measure lymphocyte blast transformation. Delayed hypersensitivity skin testing is frequently employed clinically to evaluate cell-mediated immunity in vivo.

Phagocytic Cells. Monocyte-macrophage cell lineage and polymorphonuclear leukocytes (neutrophils) are phagocytic cells that participate in immune responses and inflammation. These cells are involved in phagocytosis of microbes, immune complexes, and other foreign matter. These cells contain numerous enzymes, which are capable of degrading the phagocytized materials. Macrophages are also involved in processing antigens and presenting them to lymphocytes to initiate an immune response.

Vitamin C

Clinical Pharmacology, Metabolism, Distribution,
and Physiologic Function

In the 1930s a hexuronic acid was isolated from citrus products. Its chemical structure was defined and it was synthesized. Because of its antiscorbutic properties, this substance was called ascorbic acid [*Wilson*, 1975; *Tolbert*, 1979].

L-Ascorbic acid, or vitamin C, is passively absorbed from the gastrointestinal tract. Peak absorption occurs after 4 h. Peptic ulcer disease, diarrhea, or other intestinal disorders alter vitamin C absorption. Vitamin C is distributed in all body tissues with leukocytes considered as a labile storage pool. Normal plasma values range from approximately 5–20 µg/ml. When plasma levels exceed 14–19 µg/ml, excess absorbed vitamin C is excreted in urine. Vitamin C concentrations in leukocytes may reach 60–70 μg per 10^8 cells [*Wilson*, 1975; *Tolbert*, 1979]. In certain experimental animals vitamin C is oxidized to and excreted as CO_2 in breath. In humans, l-ascorbic acid also appears to be catabolized through dehydroascorbic acid largely to urinary metabolites [*Tolbert*, 1979; *Hornig*, 1975]. Since humans, other primates, and guinea pigs cannot convert l-gluconolactone to l-ascorbic acid, they cannot synthesize vitamin C from glucose and require the vitamin exogenously. The recommended daily allowances for vitamin C ranges from 35 (for infants) to 80 (lactating women) mg/day.

Suggested physiologic roles for vitamin C are (1) as a cofactor and activator for prolyl hydroxylase in synthesis of collagen, (2) as a cofactor in synthesis of catecholamines from tyrosine, (3) as a regulator of synthesis of steroid in gonads and adrenals, (4) as a general water-soluble reduction-oxidation agent, (5) as a growth activator in wounds and neonatal tissue, (6) as a modulator of the hexose monophosphate shunt, and (7) as an inactivator of hepatic microsomal hydroxylases [*Tolbert*, 1979]. A clear, unifying role for vitamin C in human physiology is not apparent. Deficiency of this vitamin leads to clinical scurvy.

Vitamin C possesses few pharmacological actions. It alleviates the symptoms of scurvy. Many reports have accumulated regarding effects of vitamin C, often in doses exceeding recommended daily allowance, on a wide variety of functions. Unfortunately, many of these reports are uncritical and are therefore difficult to assess. Since many of these are beyond the scope of this review, they will be mentioned only briefly. Vitamin C has been used to treat: hyperlipoproteinemia [*Ginter* et al., 1979; *Hanck and Weiser*, 1979; *Heine and Norden*, 1979; *Wilson*, 1979; *Fidanza* et al., 1982; *Ginter* et al., 1982]; cancer [*Cameron and Pauling*, 1976, 1978; *Comroe*, 1978; *Cameron* et al., 1979; *Creagan* et al., 1979; *Pauling*, 1979; *Cameron*, 1982; *Murata* et al., 1982]; infections such as upper respiratory illness [*Pauling*, 1979; *Chalmers*, 1975; *Dykes and Meier*, 1975; *Karlowski* et al., 1975; *Waldman* et al., 1975; *Coulehan* et al., 1976; Editorial, 1976; *Thomas and Holt*, 1978; *Pitt and Costrini*, 1979]; rubella [*Waldman* et al., 1975]; meningitis [*Destro and Sharma*, 1977] and post-transfusion hepatitis [*Banic and Kosac*, 1979; *Knodell* et al., 1981]; geriatric disorders [*Schorah* et al., 1979, 1981]; migraine [*Bali and Callaway*, 1978]; rheumatic fever [*Kaiser*, 1938; *Massell* et al., 1950]; rheumatoid arthritis [*Hall* et al., 1939; *Rhinehart*, 1935]; pyorrhea; allergic disorders; anemia; caries; and others. Evidence to justify use of vitamin C except in recommended daily allowances or to prevent scurvy is not convincing. The interested reader is referred to several symposia reviewing, in part, studies of vitamin C for these circumstances [*King and Burns,* 1975; *Hanck and Ritzel,* 1979; *Gross and Newberne,* 1980; *Beisel,* 1982; *Hanck,* 1982].

Immediate Hypersensitivity

Effects of vitamin C on immediate hypersensitivity have been studied in experimental conditions and in humans. *Hitchcock* [1980] found reduced release of histamine and slow-reacting substance of anaphylaxis from sensitized lung fragments of guinea pigs who were ascorbic acid-deficient; dietary replenishment of ascorbic acid restored normal in vitro responsiveness. *Gonzalez* et al. [1979] observed very weak inhibitory effects of vitamin C on degranulation and histamine release from mast cells. Similarly, *Alvarez and Mesa* [1981] confirmed little or no inhibition of rat mast cell histamine release or degranulation by ascorbic acid; at both concentrations studied (10^{-2} and 10^{-3} M) enhancement was found. Although adequate amounts of vitamin C may be necessary for appropriate experimental immediate hypersensitivity responsiveness, there is no evidence that pharmacologic doses of vitamin C ameliorate allergic responses. It has been suggested that effects of vitamin C on calcium chelation, cyclic nucleotides, phosphodiesterase, and prostaglandins [*Sharma and Wilson,* 1980] may be important in maintaining the integrity of mast cell or basophil function.

In clinical studies *Zuskin* et al. [1973] found that 500 mg ascorbic acid decreased histamine-provoked bronchoconstriction in humans and in guinea pigs. In guinea pigs, but not humans, the effect could be blocked by propranolol. This suggested a direct effect of ascorbic acid on human smooth muscle, but a β-receptor-mediated effect in guinea pigs. *Ogilvy* et al. [1981] noted that in healthy subjects ascorbic acid (1 g) reduced the bronchoconstriction that could be induced by methacholine aerosol and that this effect could be blocked by indomethacin; they suggested ascorbic acid might alter production of a bronchodilating prostaglandin. However, *Kordansky* et al. [1979] reported that in a double-blind randomized study, 500 mg vitamin C was ineffective against ragweed antigen-induced bronchospasm in known asthmatics. *Fortner* et al. [1982] showed no effect of vitamin C ingestion (2 g/day) on cutaneous or nasal reactivities to allergens in symptomatic patients. Ascorbic acid may have β-receptor or prostaglandin-mediated bronchodilatory effects on normal smooth muscles; these observations require confirmation and further consideration. Certainly, the available information is insufficient to conclude that ingestion of vitamin C would be clinically helpful for patients with allergic diatheses.

Humoral Immunity

A number of studies have examined effects of vitamin C on antibody responses of experimental animals. These indicated that ascorbate incubated with immunizing antigen enhanced antibody production [*Bourne,* 1949], that scorbutic guinea pigs had appropriate primary and secondary antibody responses [*Kumar and Axelrod,* 1969], that ascorbate-supplemented guinea pigs had a heightened IgM anti-sheep red blood cell (SRBC) response [*Prinz* et al., 1980], that chickens fed megadose vitamin C had increased antibodies to brucella [*McCorkle,* 1980], and that ascorbic acid was without effect on antibody (plaque)-forming cell responses in mice [*Siegel and Morton,* 1977]. In contrast, *Thurman and Goldstein* [1979] reported diminished cellular antibody production by scorbutic guinea pigs.

Prinz et al. [1979] observed that in humans ingestion of vitamin C, 1 g/day for 11 weeks, led to significant increases in serum IgA, IgM, and C3; but these 'significant' changes reflected in part decreases in levels of the control group. *Vallance* [1977] found IgG levels, but not IgM or C3, increased significantly in volunteers consuming 1 g/day of ascorbic acid for 1 week. *Anderson* et al. [1980b] and *Anderson and Van Wyck* [1980] found no alteration in serum IgG, IgA, IgM, C3, C4, and salivary IgA in subjects given 1–3 g of ascorbate per day for 3 weeks. Our own studies found no effect of 1 week's ingestion of 10 g/day of ascorbic acid on serum IgG, IgA, or IgM [*Panush* et al., 1982]. Effects of vitamin C on human antibody production are inconstant. Certainly there is no evidence that any of the reported changes reflect heightened host resistance or clinically important improvement in immunologic responsiveness.

Cellular Immunity

Several workers have documented that cellular immune responses in animals are dependent upon adequate ascorbate stores. *Mueller and Kies* [1962] found reduced responsiveness to mycobacteria in complete Freund's adjuvant in scorbutic guinea pigs, which was reversed by dietary ascorbate. Further, *Mueller* et al. [1962] noted that experimental allergic encephalomyelitis (a cell-mediated disease process) was induced less well in scorbutic than in normal guinea pigs. *Zweiman* et al. [1966b] found that anergic, scorbutic guinea pigs had lymphocytes that transferred delayed hypersensitivity to normal animals; lymphocytes from sensitized normal animals could not transfer immunity to scorbutic hosts. They also observed that scorbutic lymphocytes were normally responsive in vitro to tuberculin antigen to the phytomitogen phytohemagglutinin [*Zweiman* et al., 1966a] but that the inflammatory response of scorbutic animals to a nonspecific irritant was decreased [*Zweiman* et al., 1966b]. This suggested that impaired cellular responses in scorbutic animals may reflect, in part, defects in the microvasculature permitting localization of reactive cells or dysfunction of cell migration. Other investigators have documented that skin allograft survival is prolonged in scorbutic guinea pigs [*Kalden and Guthy*, 1972], presumably reflecting similar defects in ability to mount a cell-mediated response.

More recent studies in animals have confirmed and extended previous observations. Investigators have noted depressed in vitro mononuclear cell responsiveness to mitogens from vitamin C-deprived rhesus monkeys [*Hsu*, 1977] and reduced cell-mediated cytotoxicity [*Anthony* et al., 1978], delayed hypersensitivity and mitogenic responsiveness [*Thurman and Goldstein*, 1979], and T cell lymphopenia [*Fraser* et al., 1980] in scorbutic guinea pigs. Vitamin C supplementation enhanced concanavalin A responsiveness of BALB/c mouse mononuclear cells [*Siegel and Morton*, 1977] and had concentration-dependent enhancing or inhibitory effects on mouse tritiated thymidine uptake [*Tsien* et al., 1980].

Studies in humans have been undertaken during the past several years. Although some investigators described inhibition of in vitro mononuclear cell responsiveness by vitamin C [*Munster* et al., 1977; *Ramirez* et al., 1980] or no effect [*Anderson* et al., 1979], others reproducibly documented enhanced mononuclear cell responses in vitro by vitamin C [*Delafuente and Panush*, 1978; *Hahn and O'Connor*, 1978; *Panush and Delafuente*, 1979; *Manzella and Roberts*, 1979; *Delafuente and Panush*, 1980; *Panush* et al., 1982]. We were the first [*Delafuente and Panush*, 1978] to report that under specific experimental conditions, vitamin C and congeners at physiologic concentrations enhanced in vitro human mononuclear cell DNA and protein synthesis in response to stimulation by specific antigens, phytomitogens, and allogeneic cells [*Panush and Delafuente*, 1979; *Delafuente and Panush*, 1980]. *Manzella and Roberts* [1979], in an interesting set of experiments, noted that influenza virus depressed mononuclear cell responses to phytohemagglutinin, that ascorbic acid and hyperthermia counteracted the viral depressive effect, and that this action appeared macrophage-mediated. We extended our earlier observations and found that vitamin C at physiologic concentrations augmented several in vitro responses by human mononuclear cells in culture, including proliferation of cultured cell lines, function of suppressor cells, and function of natural killer cells [*Panush* et al., 1982]. We considered several possible mechanisms for the immunoenhancing effect of vitamin C. Vitamin C did not induce in vitro cellular release of soluble mediator(s) (which in turn enhanced responses of mononuclear cells), did not alter mitogen binding or receptors for phytomitogen on cells, did not exert solely nutritional effects on cell growth, did not selectively affect monocytes in our experiments, and did not abrogate suppressor cell function or signals. Rather, vitamin C appeared to directly affect cells, possibly by perturbing cellular cyclic nucleotides [*Panush* et al., 1982]. In recent experiments [*Panush* et al., 1983], we documented that vitamin C-induced increases in cyclic guanosine monophosphate (cGMP) levels in mononuclear cells, T cells, B cells, and cultured cell lines. Ascorbate-induced increased cellular cGMP had been previously noted in leukocytes, monocytes [*Sandler* et al., 1975a, b], mononuclear cells [*Gallin* et al., 1978; *Atkinson* et al., 1979], and platelets [*Schoepflin* et al., 1978; *Pickett* et al., 1979].

In the past few years there have been several attempts to determine the effect of withholding or supplementing oral vitamin C on cell-mediated immune responses of humans. Subjects with experimentally induced or spontaneous scurvy were found to have comparable numbers and functions of T cells, whether stores of vitamins C were normal or depleted [*Kay* et al., 1982]. However, normal subjects ingesting supplementary vitamin C had enhancement of certain cellular responses. *Anderson* et al. [1980a, b] recorded augmented in vitro mononuclear cell responsiveness to phytohemagglutinin and concanavalin A after subjects ingested vitamin C, 1 g/day for 1 week, 2 g/day for the second week, and 3 g/day for the third week. Our own series of studies were first reported in 1979, reflecting our effort to determine the clinical relevance, if any, of in vitro immunoenhancing properties of vitamin C. We found that normal subjects consuming 1 or 3 g, but not 10 g, of vitamin C daily for 7 days exhibited increased in vitro cellular responsiveness compared with prestudy baseline and with a placebo control group [*Panush and Delafuente,* 1979]. We subsequently reported that 2 brothers with Chediak-Higashi syndrome and defective natural killer cell function improved this function to normal levels after ingesting 6–8 g of vitamin C for 5 days [*Panush* et al., 1982]. Finally, we were able to demonstrate that vitamin C, 10 g/day for 7 days, resulted in enhanced in vivo cell-mediated immune reactivity that could be quantified. Normal subjects ingesting vitamin C exhibited increased cutaneous delayed hypersensitivity reactions to candidal and streptococcal antigens; furthermore, antigens solubilized in vitamin C induced greater reactivity than antigens in saline [*Panush* et al., 1982].

Our observations provided an interesting model for examining pharmacologic immunomodulation, suggested certain insights into the possible mechanism(s) of enhancement of the human immune response by vitamin C, and described circumstances wherein vitamin C affected cell-mediated immunity in health and disease. None of the studies reported has shown that these effects of vitamin C are useful in ameliorating disease or improving or maintaining health. We are now performing a double-blind, placebo-controlled study of the effect of vitamin C supplementation on clinical and immunologic manifestations of infectious mononucleosis, which is characterized by immune abnormalities and may be sensitive to vitamin C.

In addition to the data of *Manzella and Roberts* [1979], effects of vitamin C on monocytes/macrophages have been reported. *Ganguly* et al. [1976] found macrophages from scorbutic guinea pigs to be small and have reduced motility; the latter property was partially reversed by vitamin C. *Goetzl* et al. [1974a, b] and *Sandler* et al. [1975a] observed enhanced in vitro macrophage motility and cGMP levels [*Sandler* et al., 1975a, b] by vitamin C, and *Thomas and Holt* [1978] reported ascorbic concentration-dependent augmentation of macrophage phagocytosis.

Several studies indicated that vitamin C enhanced cellular interferon production. *Siegel* [1974] first showed that in BALB/cj mice with supplemental vitamin C in their drinking water, circulating interferon increased if stimulated with murine leukemia virus. This observation was extended to mouse L cells cultured in vitro with vitamin C and stimulated with poly-inosinic-polycytidylic acid [*Siegel,* 1975]. *Dahl and Degre* [1976] extended these studies to human embryo skin and lung fibroblasts induced by Newcastle disease virus and by polyinosinic-polycytidylic acid; *Karpinska* et al. [1982] reported similar results with human embryonic fibroblasts.

Leukocyte Functions

In addition to its effects on mononuclear cell functions and cell-mediated immune process, vitamin C affects polymorphonuclear cells in vitro and in vivo. It has been recognized for some time that appropriate phagocytic activity in guinea pigs depends on an adequate level of ascorbic acid [*Nungester and Ames,* 1948]. *DeChatelet* et al. [1972] observed that ascorbate or dehydroascorbic acid increased hexose monophosphate shunt activity in neutrophils; this was thought to occur through a sequence involving reactions of dehydroascorbate, reduced glutathione, and reduced nicotinamide adenine dinucleotide phosphate. These workers also noted high levels of ascorbic acid in phagocytic leukocytes – guinea pig peritoneal macrophages and mononuclear cells, rabbit alveolar macrophages, and human neutrophils [*DeChatelet* et al., 1974]. *Stankova* et al. [1975] reported that human neutrophil ascorbate and dehydroascorbate levels fell during phagocytosis, which are also suggestive of direct involvement of ascorbate in this process. *Stankova* et al. [1975] found that scorbutic guinea pig neutrophils phagocytized and killed as well as normals, contradicting earlier experiments of *Nungester and Ames* [1948]; this may have reflected varying results with cell populations of varying purity. *Shilotri* [1977], however, reported impaired reduced hexose monophosphate shunt activity, particle uptake and destruction, and bactericidal activity in scorbutic guinea pigs. *Goetzl* et al. [1974a, b] described additional effects of ascorbic acid on leukocyte functions. They found in vitro ascorbate enhances both random migration and chemotaxis of human neutrophils, eosinophils, and mononuclear leukocytes, possibly by effects on the hexose monophosphate shunt pathway. *Goetzl* [1976] subsequently reported decreased chemotaxis in neutrophils from patients with systemic lupus erythematosus; neutrophils from these and patients with Felty's syndrome did not show enhanced responsiveness to ascorbic acid. *Smith* et al. [1975] reported that vitamin C (5–20 mM) modestly enhanced the ability of human neutrophils to phagocytize *Candida* but inhibited intracellular killing.

Other in vitro studies of human neutrophils have indicated enhancing effects of ascorbic acid on phagocytosis and related functions. *Olson and Polk* [1977] found that ascorbic acid abrogated an inhibition of phagocytosis by corticosteroids in vitro. *Boxer* et al. [1979a, b] noted that in vitro (and in vivo) ascorbic acid promoted microtubule assembly, normalized chemotaxis and degranulation, and corrected granulocyte adherence in neutrophils from patients with Chediak-Higashi syndrome. *Anderson* and associates [1979, 1980a, b 1982] conducted a series of studies on in vitro effects of ascorbic acid on human neutrophils. They reported dose-related inhibition of *Candida* phagocytosis at supraphysiologic ascorbate concentrations (10^{-2} M) and enhancement of hexose monophosphate shunt and nitroblue tetrazolium reduction (10^{-5} to 10^{-2} M) [*Anderson,* 1979]. They corrected defective chemotaxis of neutrophils from tuberculosis patients with in vitro ascorbate (10^{-1} M) [*Gatner and Anderson,* 1980]. They confirmed [*Goetzl* et al., 1974a, b] that ascorbate stimulated neutrophil migration and reported that it inhibited (myelo) peroxidase/H_2O_2/halide system in vitro [*Anderson and Jones,* 1982]. In other studies *Kraut* et al. [1980] found ascorbate in vitro prevented an expected rise in chemiluminescence response. *Patrone* et al. [1982a] documented that ascorbic acid in vitro improved hexose monophosphate shunt activity and bacterial killing of patients with chronic granulomatous disease.

Based on these in vitro data, a number of investigators examined the effects of vitamin C on neutrophils from healthy individuals and those from patients with certain diseases. Using leukocytes from normal subjects given vitamin C, 200 mg/day for 15 days followed by 2 g/day for 2 weeks, *Shilotri and Bhat* [1977] found stimulated hexose monophosphate shunt activity at both dosage levels and impaired bactericidal activity at 2 g/day vitamin C; normal functions returned 4 weeks after vitamin withdrawal. *Anderson* [1981b, 1982] administered 1 g ascorbate intravenously to 6 volunteers and examined neutrophils 1 h later, finding slightly increased hexose monophosphate shunt activity and increased motility and decreased myeloperoxidase-mediated iodination of *Candida. Anderson* et al. [1980b] previously noted similar effects of 2–3 g ascorbate ingested daily by normal volunteers.

Although in vitro or in vivo supplementation of vitamin C tended to promote motility but not necessarily phagocytic function of normal human neutrophils, ascorbate administration did ameliorate defective neutrophil function and clinical status in certain disease states. *Chretien and Garagusi* [1973] first noted that ascorbic acid, 1 g twice a day, normalized nitroblue tetrazolium reduction in 6 patients on corticosteroid therapy (ranging from 24 mg dexamethasone to 60 mg prednisone per day, for 1 day, to more than 5 years). *Boxer* et al. [1976] used ascorbate, 200 mg/day to treat an 11-month-old with Chediak-Higashi syndrome, which is characterized in part by defective neutrophil functions. After 4 months on therapy, the child had improved neutrophil chemotaxis, improved bactericidal activity, and reduction of cellular cAMP levels; when therapy was stopped, defective neutrophil activities returned. *Gallin* et al. [1979] administered ascorbic acid, 6 g/day for 8 months, to 2 young adult brothers with Chediak-Higashi syndrome (whose natural killer cell function we subsequently examined before and during vitamin C therapy [*Panush* et al., 1982]). They noted no changes in clinical status, white blood cell counts or morphology, skin window responses, chemotaxis, bactericidal function, mononuclear cell responsiveness, or cellular cyclic nucleotide levels. They did report that administration of ascorbic acid to Chediak (beige) mice improved several neutrophil functions and increased survival. *Weening* et al. [1981] also described a teenaged patient with Chediak-Higashi syndrome who was treated with ascorbic acid, 500 mg/day for 3 years. They found ascorbate therapy decreased cellular AMP, normalized neutrophil bactericidal activity, and reduced the number of days of clinical illness. Another patient (5 months old) with this disorder, in the accelerated or lymphoma-like phase, was treated with ascorbic acid, 200 mg/day [*Saitoh* et al., 1981]; this patient manifested improved chemotaxis and bactericidal function, but the clinical course of disease was unaffected and progressed to death of the child at age 23 months. Several patients with chronic granulomatous disease have also been treated with vitamin C supplementation. *Anderson* [1981a, 1982] reported that sodium ascorbate, 1 g/day, led to increased neutrophil hexose monophosphate shunt activity, staphylocidal activity and clinical improvement in three siblings with chronic granulomatous disease, elevated IgE, and defective neutrophil motility. *Anderson* [1982] further reported improved motility of neutrophils from some patients with asthma treated with 1 g ascorbate daily. *Friedenberg* et al. [1979] also noted that ascorbic acid, 1 g 3 times a day, induced improvement – reduced rash and improved mononuclear cell and neutrophil functions in vitro – in a patient with hyperimmunoglobulin E syndrome. *Anderson and Theron* [1979] also observed improved neutrophil motility and clinical benefit in 5 of 6 patients with hyperimmunoglobulin E syndrome or recurrent bacterial infection associated with abnormal in vitro neutrophil motility. *Rebora* and colleagues [1980a, b] described several patients with neutrophil dysfunction and recurrent infection who responded to ascorbic acid supplementation (1–2 g/day); improvement of chemotaxis, killing of *Staphylococcus aureus,* and clinical benefit were noted in 5 patients. These workers recently reported similar benefit in 7 of 8 patients with recurrent skin, ear, or respiratory infections treated with ascorbate [*Patrone* et al., 1982b]. *Corberand* et al. [1982] also reported that vitamin C administration improved chemotaxis and clinical disease in a patient with *Pseudomonas* infection of the external ear.

These represent interesting but few and largely uncontrolled observations suggesting that vitamin C supplementation may be of clinical value to some patients with defective neutrophil function and resultant pyogenic problems.

Adverse Effects of Vitamin C Supplementation

The dietary allowance of vitamin C recommended by the Food and Nutrition Board of the National Academy of Sciences has been 60 mg daily, based on the ability of this amount to protect against scurvy and meet known nutritional needs of healthy persons. On the basis of extrapolation of human needs from those of guinea pigs, nonhuman primates, and other animals, this recommendation has been questioned [*Pauling,* 1970, 1974] but not altered [*Jukes,* 1975].

Consumption of vitamin C at higher levels than recommended is usually well tolerated, but may uncommonly be associated with adverse effects, including diarrhea; hemolysis in patients with glucose-6-phosphate dehydrogenase deficiency or sickle cell disease; uricosuria; oxalate and urate calculi; decreased availability of vitamin B_{12}; hypervitaminosis C after cessation of vitamin C; effects on prothrombin time during anticoagulation; possible fetal effects; effects on drug metabolism; and possibly others [*Campbell* et al., 1975; *Stein* et al., 1976; *Fulghum,* 1977; *Thomas and Holt,* 1978; Editorial, 1979; *Vilter,* 1980]. Fortunately, these occurrences are rare.

Conclusions

Clearly, vitamin C is an essential nutrient. Humans and certain animals require exogenous vitamin C to maintain adequate stores and to prevent scurvy. Daily doses of 100–150 mg vitamin C attain tissue saturation in humans; 60 mg/day is the recommended daily allowance in the USA. In experimental animals, appropriate immediate hypersensitivity and cell-mediated immune responses were dependent on vitamin C; support for a role of vitamin C in maintaining humoral immunity or neutrophil responsiveness in animals was considerably less consistent. In animals, supplemental vitamin C did not reproducibly affect mast cell function, allergen-induced bronchoconstriction, or antibody responses, but did augment cell-mediated immune responses. Studies in humans have been similar. Clinical scurvy is rare and has not been systematically examined. Supplemental vitamin C in vitro had inconsistent effects on immediate hypersensitivity, enhanced a variety of cell-mediated immune responsiveness, and enhanced many functions of neutrophils. Clinical studies of vitamin C supplementation in normal or disease states have been few and often uncontrolled. Normal or asymptomatic persons consuming vitamin C were not protected against induced bronchospasm. Some others, but not all, had increased serum Ig or C levels, increased cutaneous delayed hypersensitivity reactions, increased cell-mediated immune responses in vitro, and increased neutrophil motility. Vitamin C has been administered with immunologic and clinical benefit to some patients with immunologically mediated diseases, including Chediak-Higashi syndrome, chronic granulomatous disease, and recurrent infections. Vitamin C has also been used to treat conditions that may, in part, be mediated by immunologic events, such as cancer, a variety of infections, rheumatic and allergic diseases, and others; but compelling evidence of efficacy is lacking. It seems evident that vitamin C is necessary to maintain certain aspects of the integrity of the immune system and host defense mechanisms. Some clinical conditions might be responsive to supplemental vitamin C, particularly some characterized by defective neutrophil function.

Vitamin E

Humoral Immunity

The effect of vitamin E on humoral immune responses has been extensively studied in animals. In one study, hens and their chicks were maintained on a diet fortified with vitamin E, 60 mg per pound of feed [*Tengerdy* et al., 1972]. 7-day-old chicks receiving this diet were immunized with SRBC. Their spleens and bursas were subsequently removed, and the antibody-secreting cells of these organs were tested for their ability to make antibody to SRBC in a plaque-forming cell assay. Chicks that received the vitamin E-supplemented diet produced significantly more plaque-forming cells than the control group receiving a regular diet. Greatest antibody enhancement was found in 1-week-old hypoxic chicks hatched at natural or simulated high altitudes. This suggested to the investigators that one possible mode of action for immunoenhancement by vitamin E was as an antioxidant or redox agent that acted synergistically with hypoxia in creating favorable reducing conditions, which stimulated immunopoietic cellular development and proliferation [*Tengerdy,* 1980].

Chicks receiving SRBC immunization at 7 days of age were reimmunized at 31 days of age [*Tengerdy* et al., 1972]. Again, spleen cells from the group receiving vitamin E-supplemented diets produced significantly more plaque-forming cells than the control group. There were also corresponding higher hemagglutination titers in the chicks on a diet high in vitamin E content. Similar enhanced antibody responses to immunization were also seen in 1-year-old hens on the vitamin E-fortified diet. These researchers also measured hemagglutination titers at weekly intervals after chicks hatched and were fed diets supplemented with 0, 30, or 60 mg of vitamin E per pound of feed. Hemagglutination titers increased with age, and the antibody responses were always greater in the vitamin E-treated group. The group receiving the largest dose of vitamin E had the greatest enhancement in antibody production.

The effect of vitamin E on humoral immunity has also been studied in guinea pigs [*Barber* et al., 1977]. Guinea pigs receiving vitamin E for 2 weeks were immunized with attenuated Venezuelan equine encephalomyelitis virus vaccine, and hemagglutination inhibition titers were subsequently measured. Intramuscular administration of vitamin E enhanced in vivo antibody production in guinea pigs. The greatest enhancement occurred in animals receiving exceedingly large doses of the vitamin (33 IU/kg). The failure of oral ingestion of vitamin E to enhance humoral immune responses in these guinea pigs may have been due to the formulation of vitamin E used. Because the acetate ester of vitamin E was used, these animals may have been unable to cleave the acetate ester and convert the vitamin to its active form.

In other studies, mice were maintained on diets supplemented with 0, 20, or 200 mg of vitamin E per kilogram of feed [*Tanaka* et al., 1979]. After receiving these diets for 50 days, the mice were immunized with foreign red blood cells (RBC). Hemagglutination titers were measured before and 4, 8, and 14 days after immunization. Antibody responses to the RBC on days 8 and 14 were significantly enhanced in the mice receiving the largest dose of vitamin E compared with the mice receiving no vitamin E in their diets. There was no enhancement of antibody production in mice receiving a standard diet containing 20 mg of vitamin E per kilogram of feed. 4 weeks after the primary RBC immunization, mice were immunized with RBC conjugated to the hapten trinitrophenyl. Vitamin E, in a dose-dependent fashion, enhanced the anti-hapten antibody response to immunization with a hapten-carrier conjugate, possibly by stimulation of carrier-specific helper T cells.

Other studies used mice that were fed diets with or without oral vitamin E supplementation before immunization with SRBC or tetanus toxoid [*Tengerdy* et al., 1973]. The vitamin E-treated group produced about 40% more plaque-forming cells to SRBC with a corresponding increase in serum hemagglutination titers. The increase in plaque-forming cells probably reflects an increase in the number of antibody-producing cells, rather than an increase in antibody secretion by a single antibody-secreting cell, because spleen weights in the vitamin E-treated group were larger. Vitamin E had a more pronounced effect on primary immune responses than on secondary responses, and IgG synthesis was affected more than IgM synthesis. The IgG response to SRBC was almost completely suppressed in mice fed a diet deficient in vitamin E, but the antibody response was restored by adding vitamin E back to the diet. Similar results were obtained using tetanus toxoid as the antigen. The antioxidant diphenyl-p-phenylene diamine could not substitute for vitamin E.

Mice were used to study the effects of vitamin E on subpopulations of cells involved in immune responses [*Campbell* et al., 1974]. Spleen cells from mice were cultured in the presence of SRBC with and without vitamin E. Plaque-forming cells were determined after 5 days of culture. As in other studies, vitamin E enhanced the antibody response to SRBC. When adherent cells (macrophages) were removed and the nonadherent cell population (lymphocytes) cultured with SRBC, vitamin E was still able to enhance plaque-forming cell responses. Normally, adherent cells are required to cooperate with the nonadherent cells for an immune response. The antioxidant 2-mercaptoethanol can replace the adherent cells in many assay systems, and it was found that 2-mercaptoethanol was as effective as vitamin E in increasing the number of plaque-forming cells in a nonadherent cell population. This suggests an apparent macrophage bypass or an enhancement of the few macrophages contaminating the nonadherent cells. Since vitamin E and 2-mercaptoethanol are both antioxidants, they may enhance immune responses by similar mechanisms. *Campbell* et al. [1974] have proposed that these agents may increase respiratory and metabolic activities of cells by keeping sulfhydryl groups on cell membranes in the reduced state. Another possible mechanism of action may involve a protective effect against oxygen toxicity. Vitamin E increased cell recovery in some of the experiments with nonadherent cells and SRBC. This may have been due to enhanced cell growth or decreased cell death, or to both.

Cellular Immunity

The effect of vitamin E on cellular immune responses has also been examined. Vitamin E at physiologic concentrations enhanced blast transformation of mouse spleen lymphocytes stimulated by suboptimal concentrations of concanavalin A. These effects of vitamin E occurred in the presence and absence of adherent cells [*Corwin and Schloss,* 1980]. These investigators also examined the in vivo effects of vitamin E on cell-mediated immunity. Mice were fed a diet containing 0, 5, or 50 mg of vitamin E per 100 g of feed. Both the 5- and 50-mg vitamin E supplements enhanced spleen cell responses to concanavalin A. Responses to phytohemagglutinin were enhanced only in mice receiving the diet containing 50 mg of vitamin E per 100 g of feed. Similar but less dramatic effects occurred when lipopolysaccharide was used as the lymphocyte stimulator. Concanavalin A, phytohemagglutinin, and lipopolysaccharide are known to stimulate different subpopulations of lymphocytes. Because the effects of vitamin E varied with the lymphocyte stimulus used, it would appear that vitamin E may have different effects on different lymphocyte subpopulations.

The enhancement of lymphocyte blastogenesis by vitamin E may involve an interaction between the vitamin and the thymus or T lymphocytes. When lymphocytes from congenitally athymic mice were cultured with lipopolysaccharide (B cell mitogen) and vitamin E, the responses were less than those from mice with normal thymuses [*Corwin and Schloss,* 1980]. These data suggest that vitamin E may act on helper T cells to augment lymphocyte responses. The enhancement of lymphocyte responses by vitamin E has not been a consistent finding. In vitro responses to phytohemagglutinin stimulation by peripheral blood lymphocytes from healthy human subjects were significantly decreased after ingestion of 300 mg of vitamin E daily for 3 weeks [*Prasad,* 1980]. Vitamin E ingestion had no effect on delayed cutaneous hypersensitivity reactions to phytohemagglutinin in this healthy group. Unfortunately, these studies did not contain a placebo-controlled group.

The effects of vitamin E on survival from bacterial infections have been studied [*Heinzerling* et al., 1974]. In these experiments mice were fed a standard diet with or without varying quantities of vitamin E for 4 weeks before challenge with live *Diplococcus pneumoniae.* Some of the animals in these studies had been previously immunized with the same organism. 5 days after the challenge, the percentage of mice surviving was higher in the vitamin E-treated mice. This phenomenon was true for both immunized and nonimmunized mice. The optimal dose of vitamin E, 180 mg/kg of feed, increased survival 4-fold above that in the controls. There was a biphasic dose-response curve of survival with varying vitamin E doses. Survival in the group receiving 360 mg/kg of feed equaled that in the 180 mg/kg group; however, intermediate doses of vitamin E were less effective for increasing survival. The degree of enhanced protection from vitamin E supplementation was similar for the immunized and nonimmunized mice; however, the challenge doses of the bacteria differed by a factor of 1,000.

Heinzerling et al. [1974] also examined the effects of vitamin E supplementation on the ability of leukocytes to phagocytize the diplococci. 4 h after being challenged with the organism, mice receiving vitamin E at doses of 120 or 300 mg/kg of feed had a 4-fold increase in the number of macrophages phagocytizing the bacteria. Macrophages from vitamin E-treated mice were also more active than controls in bacterial lysis. Again, a biphasic dose response was evident. Phagocytosis of carbon particles in the blood was also more efficient in the vitamin E-treated mice, which indicated a general nonimmunologic enhancement in phagocytic activity. These investigators speculated that vitamin E may increase ubiquinone synthesis, leading to more efficient electron transport and metabolism in phagocytic and immunocompetent cells.

There are limited human data describing the effects of vitamin E on phagocytosis and bactericidal activity. The phagocytic rate of lipopolysaccharide-red paraffin oil particles by human polymorphonuclear leukocytes has been studied [*Baehner* et al., 1977]. In these experiments 3 human subjects ingested 400 IU of vitamin E 4 times daily for 7 days. The vitamin markedly increased the neutrophil phagocytic rate. Interestingly, the bactericidal capacity of neutrophils from these subjects was decreased during vitamin E supplementation. This was attributed to less hydrogen peroxide available intracellularly due to a decrease in the rate of superoxide anion (O_2^-) generation.

Another uncontrolled study [*Prasad*, 1980] also demonstrated a significant decrease in bactericidal activity by human blood leukocytes after ingestion of 300 mg of vitamin E daily for 3 weeks. Release of acid phosphatase by phagocytizing cells was shown to be significantly decreased. Vitamin E, in a dose-dependent manner, decreased oxygen consumption by neutrophils phagocytizing bacteria [*Repine* et al., 1978]. Vitamin E also caused a partial inhibition of glucose oxidation in phagocytizing neutrophils. In these studies the vitamin did not inhibit phagocytosis. Electron microscopy showed that vitamin E had no effect on surface membranes, cytoplasmic structures, storage organelles, or nuclei of neutrophils.

Conclusions

Vitamin E has repeatedly been shown to enhance in vitro and in vivo antibody production in mice, chickens, and guinea pigs. Vitamin E ingestion increased in vitro proliferation of antibody-secreting cells. Primary antibody responses appear to be more sensitive than secondary responses to the modulating effects of vitamin E. Vitamin E can replace the requirement for macrophages in immune functional assays in vitro. These data suggest that vitamin E may interact with cooperating immunocompetent cells to evoke an immune response. Experimental results have implied that vitamin E may act on helper T lymphocytes to augment immune responses. Vitamin E has also been shown to have beneficial effects on cellular immunity in animal models. Both phagocytosis and lymphocyte proliferation are augmented by vitamin E.

Data on the effects of vitamin E on the immune response in humans are limited; animal data should not necessarily be extrapolated to humans without experimental confirmation. Two uncontrolled studies have shown that pharmacologic doses of vitamin E in vivo depress both lymphocyte proliferation and leukocyte bactericidal capacity. More studies are needed before a definitive role can be established for this vitamin in human immunity.

Group B Vitamins

Pyridoxine
Humoral Immunity
The effect of pyridoxine on the immune system has been studied extensively in animals. Early reports showed a marked effect on humoral immunity [*Axelrod,* 1958]. In these studies rats were maintained on a pyridoxine-deficient diet for 4 weeks, then immunized with diphtheria toxoid. Primary antibody responses were measured 3 weeks later, and a booster immunization was given 4 weeks after the first immunization. Both primary and secondary antibody responses were depressed in the pyridoxine-deficient rats. Rats that were restored on a diet containing pyridoxine during the second toxoid challenge did not have an anamnestic response. In other experiments [*Axelrod,* 1958] acute pyridoxine deficiency was induced during the secondary response by using the antagonist deoxypyridoxine. Although primary antibody responses were normal before induction of pyridoxine deficiency, the secondary antibody responses were abrogated. These studies emphasize the need for pyridoxine for both primary and secondary humoral immune responses. If the vitamin is absent during a primary immunization, anamnestic responses will not occur even when the nutritional deficiency is restored.

Guinea pigs have also been used to study the effects of pyridoxine on immune responses [*Axelrod* et al., 1961]. These animals were immunized with diphtheria toxoid after receiving the antagonist deoxypyridoxine and a pyridoxine-deficient diet. Primary antibody responses were diminished compared with those in nonpyridoxine-deficient guinea pigs. Repeat immunization also failed to elicit a significant secondary antibody response in the deficient animals. Deficient animals also did not have a local reaction to intradermal testing with the toxoid, whereas guinea pigs receiving a normal diet did.

Cellular Immunity

The effect of pyridoxine deficiency on delayed hypersensitivity reactions has also been examined [*Axelrod* et al., 1963]. Guinea pigs were sensitized with *Mycobacterium tuberculosis.* Delayed cutaneous hypersensitivity reactions to purified protein derivative were subsequently measured. Pyridoxine-deficient guinea pigs consistently had depressed reactions. When these animals were switched to a diet containing the vitamin, skin reactivity to purified protein derivative was equal to that of controls. In addition, the mycobacterium caused fewer lesions in the omentum of the pyridoxine-deficient animals. Although the effector function of the cellular elements of the immune response was blunted, in vitro testing and passive transfer experiments demonstrated that spleen cells from the deficient animals were sensitized to purified protein derivative.

Other evidence also suggests that pyridoxine is necessary for an intact cell-mediated immune response. Rejection of skin grafts between different strains of rats is reduced in pyridoxine-deficient animals [*Axelrod* et al., 1958]. In these experiments, graft survival was 15% when both the donor and the recipient rats were maintained on a normal diet. When the donor was pyridoxine-deficient and the recipient was not, graft survival rate was 28%; the rate increased to 62% when the donor was on a normal diet and the recipient was pyridoxine-deficient. When both the recipient and the donor were pyridoxine-deficient, graft survival was 90%. These data clearly show pyridoxine deficiency can inhibit transplantation rejection, which is a function of cellular immunity.

In similar studies [*Fisher* et al., 1958], pyridoxine deficiency in both the recipient and the donor enhanced graft survival in rats. Returning the recipient animals to a normal diet 3–4 weeks after transplantation did not increase the incidence of rejection. Other evidence that pyridoxine is essential for cellular immunity is that pyridoxine-deficient guinea pigs immunized with Calmette-Guérin bacillus failed to respond normally with a delayed cutaneous reaction to intradermal purified protein derivative [*Trakatellis* et al., 1963]. These same deficient animals were able to elicit anaphylactic reactions in response to an intravenous injection of a protein antigen to which they were sensitive.

Other B Vitamins

The roles of biotin and pantothenic acid on humoral immune responses have been studied [*Axelrod,* 1958]. Rats were fed a diet deficient in one of these nutrients for 4 weeks before immunization with diphtheria toxoid. Primary antibody responses were measured after 3 weeks, then a booster immunization was given, and antibody responses were determined after 1 week. Some animals received a regular diet including the missing vitamin during the week of the secondary response. Lack of either biotin or pantothenic acid prevented normal primary and secondary humoral immune responses. Supplementation of the missing vitamin during the secondary response had little effect on antibody responses. This emphasizes the need for these nutrients during the primary immune response in order for secondary immunologic stimulation to be effective.

Numerous single vitamin deficiencies were induced in rats to examine their effects on antibody formation. Deficiencies in either pantothenic acid, pyridoxine, or folic acid caused a severe impairment in antibody synthesis to both diphtheria toxoid and human RBC [*Axelrod*, 1971]. There was a moderate inhibition in antibody responses to deficiencies in thiamin, biotin, riboflavin, or niacin. Deficiency of vitamin B_{12} did not alter the humoral immune response. Other investigators have shown a potent inhibition of primary antibody responses using the folic acid antagonist methotrexate in guinea pigs and dogs [*Friedman* et al., 1961; *Donnall* et al., 1963]. These data confirm the need for folic acid in the development of immune responses.

One study has determined the effects of a combined deficiency of pantothenic acid and pyridoxine in humans [*Hodges* et al., 1962]. In this study, 2 men received normal diets, 2 men received a diet deficient in both vitamins, and 2 men received the vitamin antagonists deoxypyridoxine and omega-methylpantothenic acid. All subjects were immunized with tetanus toxoid and typhoid vaccine after 2, 3, and 4 weeks on the diets. The men on the deficient diet had normal antibody responses, whereas the men given the vitamin antagonists had no response to the tetanus toxoid and a diminished antibody response to the typhoid vaccine.

The mechanisms of action for most of the vitamins on the immune response are not known. Pyridoxine deficiency is known to impair nucleic acid synthesis, with subsequent inhibitory effects on protein synthesis, DNA synthesis, and cell proliferation. These deleterious effects on immunocompetent cells are probably responsible for the immunologic impairment in pyridoxine deficiency. With pantothenic acid deficiency antibody producing cells in the spleen are markedly decreased in number; however, the reticuloendothelial cells process antigen normally [*Axelrod*, 1971].

Conclusions

Deficiencies of pyridoxine, pantothenic acid, or folic acid cause severe impairments in humoral- and cell-mediated immunity in many animal species. The B vitamins thiamine, biotin, riboflavin, and niacin only moderately interfere with immunologic processes. The B vitamins comprise a large number of molecules that differ markedly in chemical structure and biologic activities. Thus, it is not surprising that single deficiencies of these vitamins produce varied effects on the immune response. It is also not surprising that deficiencies of folic acid and pyridoxine cause severe abnormalities in the immune response, since these vitamins are essential for normal intracellular metabolism. Therefore, aberrations in immune responses from deficiencies in the B vitamins may not be specific to the immune system, but secondary to a global inhibition of cellular function.

The importance of supplemental B vitamins in nondeficiency states is not known. There are no data from animals or humans on the effects of supraphysiologic levels of these vitamins on immunocompetence.

Vitamin A

Humoral Immunity

Data available indicate that vitamin A can modulate humoral immune responses. Mice were injected with vitamin A 4 days before immunization with SRBC [*Cohen and Cohen,* 1973]. Spleen cells were obtained and a plaque-forming cell assay was performed 4 days after immunization. Vitamin A, in a dose-response fashion, increased the number of plaque-forming cells with doses of 3,000 IU/day, giving maximum responses that were 5-fold higher than controls. A dose of 9,000 IU/day was toxic, and there was no enhancement in plaque-forming cell. Using similar methodology, other investigators were unable to show enhancement of plaque-forming cell responses by using similar dosages of vitamin A [*Dennert and Lotan,* 1978]. These investigators also examined the effects of vitamin A on serum antibody levels to a soluble antigen. Mice received two daily injections of vitamin A before immunization. 2,000 and 5,000 IU of the vitamin significantly enhanced antibody production at 3 weeks. Low-dose hydrocortisone before immunization resulted in a decrease in plaque-forming cell responses in mice; however, concomitant vitamin A (3,000 IU) prevented this hydrocortisone-induced immunosuppression [*Cohen and Cohen,* 1973]. Vitamin A had no effect on large doses of hydrocortisone.

Similar results have been reported by other researchers. Mice received 3,000 IU of vitamin A 2 days before and the day of immunization with bovine serum albumin [*Falchuk* et al., 1977]. Vitamin A was also administered at 2-week intervals along with a booster of bovine serum albumin. Vitamin A augmented the antibovine serum albumin antibody response. In addition, there were no differences in responses from the palmitate and alcohol forms of vitamin A, or from oral or subcutaneous administration.

The time of vitamin A administration in relation to the time of antigen challenge is critical for immunoenhancement. Vitamin A injected daily for either 5 days before or after sensitization enhanced hemagglutination titers to SRBC [*Jurin and Tannock,* 1972]. Injection of vitamin A starting 6 days after immunization had no effect on antibody production.

Vitamin A-deficient rats, compared with rats supplemented with 50 μg/day of vitamin A, have depressed serum antibody responses to antigenic challenges and have decreased ability to produce plaque-forming cells in vitro [*Krishnan* et al., 1974]. Part of this impairment was due to inanition, since rats receiving 50 μg/day of vitamin A and a low calorie diet, which maintained their weight loss equivalent to the vitamin A-deficient group, also had depressed humoral responses.

Children in Bangladesh were used to study the effects of vitamin A on human immune responses [*Brown* et al., 1980]. These children had never received tetanus toxoid vaccination and had negative skin tests to the toxoid. The children were immunized with tetanus toxoid, with or without 200,000 IU of vitamin A injected intramuscularly at the time of vaccination. 4 weeks after the primary immunization, neither the vitamin A group nor the control group had demonstrable antitetanus toxoid antibodies. 8 weeks after a booster immunization, there was no difference in antitetanus toxoid antibodies or skin test reactivity between the two groups. Although none of these children were vitamin A-deficient, they may have been undernourished, since their weights were less than predicted for their corresponding ages.

Vitamin A has been shown to enhance survival from infections with a variety of microorganisms. Mice received 3,000 IU of vitamin A daily for 4 days before becoming infected [*Cohen and Elin,* 1974]. In animals infected with *Pseudomonas aeruginosa,* vitamin A significantly enhanced survival. For the first 3 h after inoculation, both the vitamin A and control groups had equivalent bacteremias. However, no organisms could be cultured in the vitamin A group 5 h after becoming infected. After 24 h all control mice died, whereas the vitamin A group was still sterile. Control mice infected with *Candida albicans* had a 65% mortality rate, but there was only a 5% mortality rate in vitamin A-treated mice. Similar effects were seen with *Listeria monocytogenes*. The enhancing effects of vitamin A seemed to be independent of phagocytosis because clearance of carbon particles and aggregated albumin from the blood were not altered by the vitamin.

Cellular Immunity

Rats were maintained on a vitamin A-deficient diet to examine the effects on cell-mediated immunity [*Nauss* et al., 1979]. Spleen cell responses to concanavalin A, phytohemagglutinin, and lipopolysaccharide were significantly diminished compared with those in rats receiving 500 µg of vitamin A per week. Although the vitamin A-deficient rats lost significant body weight, inanition did not account for the blunted immune responses. Pair-fed rats that lost equal weight but were supplemented with 500 µg of vitamin A per week had normal spleen cell mitogenic responses. When vitamin A was restored to the diet, mitogenic responses returned to normal within 3 days. In these studies, vitamin A deficiency did not alter thymus cell responses to concanavalin A.

Skin graft rejection in mice was accelerated by the administration of vitamin A [*Jurin and Tannock,* 1972]. Mean rejection time decreased from 35.5 to 23.7 days in vitamin A-treated mice. Other investigators reported similar findings [*Medawar and Hunt,* 1981]. Daily vitamin A supplementation decreased the mean survival time skin grafts by 7.5 days. Giving the recipient vitamin A for 14 days before transplantation decreased survival time by 11 days. This augmentation of graft rejection could be abrogated by treating the mice with antithymocyte serum, which suggests that vitamin A modulates T lymphocytes.

Vitamin A has also been shown to enhance survival from tumors. Mice were supplemented with vitamin A orally (approximately 3,100 IU/day) for 2 weeks before injection of melanoma cells [*Felix* et al., 1975]. All control mice developed tumor; 17% of treated mice developed tumor. In addition, the size of developing tumor in the treated mice lagged behind that in the controls. The tumor protection from vitamin A could be overcome by anti-lymphocyte serum, indicating the antitumor effect of the vitamin is at least partially due to enhancement of lymphocytes. The antitumor effect of vitamin A appears to depend upon the type of tumor. Vitamin A had no effect on tumor growth for 6 different transplantable tumors [*Bollag,* 1971]. Chemically induced skin papillomas regressed by 61% in mice receiving

100 mg/kg/week of vitamin A. Tumor regression increased to 86% when the dosage of vitamin A was increased to 400 mg/kg/week [*Bollag,* 1971].

Mice were treated with varying dosages of vitamin A for 7 days before being injected with tumor cells [*Dennert and Lotan,* 1978]. After another 7 days spleen cells from these mice were tested for tumor-specific cytotoxic activity. Vitamin A at doses of 25 and 100 µg/day increased cytotoxic activity 10-fold compared with that in controls. Although 300 µg/day of the vitamin enhanced cytotoxicity, it was less than that with lower doses. Higher doses either had either no effect or suppressed cell-mediated cytotoxicity. In vitro sensitization of spleen cells to tumor cells in the presence of vitamin A also generated up to 10-fold higher cytotoxic activity than that in controls. This effect occurred in a dose-response fashion with optimal concentrations at 10^{-8} to 10^{-6} M.

The effect of vitamin A on lymphoid organs has also been examined. Mice received daily vitamin A supplementation, 250 mg/kg, for 21 days [*Medawar and Hunt,* 1981]. There was a slight, but significant, decrease in total body weight in treated mice. Spleen weights were not different between control and treated mice. There were significant increases in lymph node and thymus weights in the vitamin A-treated mice. Histologic examination revealed hypertrophy of peripheral lymphoid organs with a relative enlargement of the paracortical areas in treated mice. There was also an absolute increase in the number of lymph node cells bearing the theta antigen (T lymphocytes).

Vitamin A deficiency in rats resulted in atrophy of spleens and thymuses [*Krishnan* et al., 1974]. Inanition accounted for a portion of the atrophy. Thymus cortices from vitamin A-deficient animals were almost completely depleted of lymphocytes, whereas the medulla appeared to be unaffected. There was also marked involution of spleen germinal centers in deficient animals.

Conclusions

Vitamin A was capable of modifying immune responses, both in vivo and in vitro, in a variety of animal models. Vitamin A had adjuvant properties when administered at the time of antigenic challenge. The mechanisms for this action may be related to increased labilization of lysosomal membranes, increased lymphocyte proliferation, or altered macrophage antigen processing [*Falchuk* et al., 1977; *Cohen and Cohen,* 1973]. In addition to enhancing antibody production, vitamin A augmented cellular immune responses, as shown by accelerated transplantation rejection, enhanced mitogenic responses, and increased cell-mediated cytotoxicity. Vitamin A enhanced guinea pig peritoneal macrophage production of arginase, a tumoricidal enzyme [*Rhodes and Oliver,* 1980], which may account for tumor regression and increased host survival. A role for vitamin A in human immunity has not been clearly identified.

References

Alvarez, R.G.; Mesa, M.G.: Ascorbic acid and pyridoxine in experimental anaphylaxis. Agents Actions *11:* 89–93 (1981).

Anderson, R.: Effects of ascorbate on leucocytes. II. Effects of ascorbic acid and calcium and sodium ascorbate on neutrophil phagocytosis and post-phagocytosis and post-phagocytic metabolic activity. S. Afr. med. J. *56:* 401–404 (1979).

Anderson, R.: Assessment of oral ascorbate in three children with chronic granulomatous disease and defective neutrophil motility over a 2-year period. Clin. exp. Immunol. *43:* 180–188 (1981a).

Anderson, R.: Ascorbate-mediated stimulation of neutrophil motility and lymphocyte transformation by inhibition of the peroxidase/H_2O_2/halide system in vitro and in vivo. Am. J. clin. Nutr. *34:* 1906–1911 (1981b).

Anderson, R.: Effects of ascorbate on normal and abnormal leukocyte functions. Int. J. Vitam. Nutr. Res. *23:* suppl., pp. 23–34 (1982).

Anderson, R.; Hay, I.; Van Wyk, H.; Oosthuizen, R.; Theron, A.: The effect of ascorbate on cellular and humoral immunity in asthmatic children. S. Afr. med. J. *58:* 974–977 (1980a).

Anderson, R.; Jones, P.T.: Increased leucoattractant binding and reversible inhibition of neutrophil motility mediated by the peroxidase/H_2O_2/halide system: effects of ascorbate, cysteine, dithiothreitol, levamisole and thiamine. Clin. exp. Immunol. *47:* 487–496 (1982).

Anderson, R.; Oosthuizen, R.; Gatner, E.M.: Effects of ascorbate on leukocytes. V. Effects of ascorbate and calcium and sodium ascorbate on certain functions of human blood lymphocytes in vitro. S. Afr. med. J. *56:* 511–515 (1979).

Anderson, R.; Oosthuizen, R.; Maritz, R.; Theron, A.; Van Rensburg, A.J.: The effects of increasing weekly doses of ascorbate on certain cellular and humoral immune functions in normal volunteers. Am. J. clin. Nutr. *33:* 71–76 (1980b).

Anderson, R.; Theron, A.: Effects of ascorbate on leucocytes. III. In vitro and in vivo stimulation of abnormal neutrophil motility by ascorbate. S. Afr. med. J. *56:* 429–433 (1979).

Anderson, R.; Van Wyk, H.: The effects of ascorbate ingestion on levels of salivary IgA in adult volunteers (Letter). Am. J. clin. Nutr. *33:* 6–7 (1980).

Anthony, L.E.; Kurahara, C.G.; Taylor, K.B.: Immunocompetence and ascorbic-acid deficiency in guinea-pigs (Abstract). Fed. Proc. *37:* 931 (1978).

Atkinson, J.P.; Weiss, A.; Ito, M.; Kelly, J.; Parker, C.W.: Effects of ascorbate on cyclic nucleotide metabolism in human lymphocytes. J. cyclic Nucleotide Res. *5:* 107–123 (1979).

Axelrod, A.E.: The role of nutritional factors in the antibody responses of the anamnestic response. Am. J. clin. Nutr. *6:* 119–125 (1958).

Axelrod, A.E.: Immune processes in vitamin deficiency states. Am. J. clin. Nutr. *24:* 265–271 (1971).

Axelrod, A.E.; Fisher, B.; Fisher, E.; Lee, C.P.; Walsh, P.: Effect of a pyridoxine deficiency on skin grafts in the rat. Science *127:* 1388–1399 (1958).

Axelrod, A.E.; Hopper, S.; Long, D.A.: Effects of pyridoxine deficiency upon circulating antibody formation and skin hypersensitivity reactions to diphtheria toxoid in guinea pigs. J. Nutr. *74:* 58–64 (1961).

Axelrod, A.E.; Trakatellis, A.C.; Bloch, H.; Stinebring, W.R.: Effect of pyridoxine deficiency upon delayed hypersensitivity in guinea pigs. J. Nutr. *79:* 161–167 (1963).

Baehner, R.L.; Boxer, L.A.; Allen, J.M.; Davis, J.: Autooxidation as basis for altered function by polymorphonuclear leukocytes. Blood *50:* 327–335 (1977).

Bali, L.; Callaway, E.: Vitamin C and migraine: a case report (Letter). New Engl. J. Med. *299:* 364 (1978).

Banic, S.; Kosak, M.: Prevention of transfusion hepatitis by vitamin C. Int. J. Vitam. Nutr. Res. *19:* suppl., pp. 41–44 (1979).

Barber, T.L.; Nockels, C.F.; Jochim, M.M.: Vitamin E enhancement of Venezuelan equine encephalomyelitis antibody response in guinea pigs. Am. J. vet. Res. *38:* 731–734 (1977).

Beisel, W.R.: Single nutrients and immunity. Am. J. clin. Nutr. *35:* suppl., pp. 417–468 (1982).

Bloch, K.J.: Antibodies and their functions; in Benacerraf, Unanue, Textbook of immunology, pp. 31–53 (Williams & Wilkins, Baltimore 1979).

Bollag, W.: Effects of vitamin A acid on transplantable and chemically induced tumors. Cancer Chemother. Rep. *55:* 53–58 (1971).

Bourne, G.H.: Vitamin C and immunity. Br. J. Nutr. *2:* 341–347 (1949).

Boxer, L.A.; Albertini, D.F.; Baehner, R.L.; Oliver, J.M.: Impaired microtubule assembly and polymorphonuclear leukocyte function in the Chediak-Higashi syndrome correctable by ascorbic acid. Br. J. Haemat. *43:* 207–213 (1979a).

Boxer, L.A.; Vanderbilt, B.; Bonsib, B.; Jersild, R.; Yang, H.-H.; Baehner, R.L.: Enhancement of chemotactic response and microtubule assembly in human leukocytes by ascorbic acid. J. cell. Physiol. *100:* 119–126 (1979b).

Boxer, L.A.; Watanabe, A.M.; Rister, M.; Besch, H.R., Jr.; Allen, J.; Baehner, R.L.: Correction of leukocyte function in Chediak-Higashi syndrome by ascorbate. New Engl. J. Med. *295:* 1041–1045 (1976).

Brown, K.H.; Rajan, M.M.; Chakraborty, J.; Aziz, K.M.A.: Failure of a large dose of vitamin A to enhance the antibody response to tetanus toxoid in children. Am. J. clin. Nutr. *33:* 212–217 (1980).

Campbell, G.D.; Steinberg, M.H.; Bower, J.D.: Ascorbic acid-induced hemolysis in G-6PD deficiency. Ann. intern. Med. *82:* 810 (1975).

Campbell, P.A.; Cooper, H.R.; Heinzerling, R.H.; Tengerdy, R.P.: Vitamin E enhances in vitro immune response by normal and nonadherent spleen cells. Proc. Soc. exp. Biol. Med. *146:* 465–469 (1974).

Cameron, E.: Vitamin C and cancer: an overview. Int. J. Vitam. Nutr. Res. *23:* 115–127 (1982).

Cameron, E.; Pauling, L.: Supplemental ascorbate in the supportive treatment of cancer: prolongation of survival times in terminal human cancer. Proc. natn. Acad. Sci. USA *73:* 3685–3689 (1976).

Cameron, E.; Pauling, L.: Supplemental ascorbate in the supportive treatment of cancer: reevaluation of prolongation of survival times in terminal human cancer. Proc. natn. Acad. Sci. USA *75:* 4538–4542 (1978).

Cameron, E.; Pauling, L.; Leibovitz, B.: Ascorbic acid and cancer: a review. Cancer Res. *39:* 663–681 (1979).

Chalmers, T.C.: Effects of ascorbic acid on the common cold. An evaluation of the evidence. Am. J. Med. *58:* 532–536 (1975).

Chretien, J.H.; Garagusi, V.F.: Correction of corticosteroid-induced defects of polymorphonuclear function by ascorbic acid. J. reticuloendoth. Soc. *14:* 280–286 (1973).

Cohen, B.E.; Cohen, I.K.: Vitamin A: adjuvant and steroid antagonist in the immune response. J. Immun. *111:* 1376–1380 (1973).

Cohen, B.E.; Elin, R.J.: Vitamin A induced nonspecific resistance to infection. J. infect. Dis. *129:* 597–600 (1974).

Comroe, J.: Experimental studies designed to evaluate the management of patient with incurable cancer. Proc. natn. Acad. Sci. USA *75:* 4543 (1978).

Corberand, J.; Nguyen, F.; Fraysse, B.; Enjalbert, L.: Malignant external otitis and polymorphonuclear leukocyte migration impairment. Improvement with ascorbic acid. Archs Otolar. *108:* 122–124 (1982).

Corwin, L.M.; Schloss, J.: Influence of vitamin E on the mitogenic response of murine lymphoid cells. J. Nutr. *110:* 916–923 (1980).

Coulehan, J.L.; Eberhard, S.; Kapner, L.; Taylor, F.; Rogers, K.; Garry, P.: Vitamin C and acute illness in Navajo school children. New Engl. J. Med. *295:* 973–977 (1976).

Creagan, E.T.; Moertel, C.G.; O'Fallon, J.R.; Schutt, A.J.; O'Connell, M.J.; Rubin, J.; Frytak, S.: Failure of high-dose vitamin C (ascorbic acid) therapy to benefit patients with advanced cancer. A controlled trial. New Engl. J. Med. *301:* 687–690 (1979).

Dahl, H.; Degre, M.: The effect of ascorbic acid on production of human interferon and the antiviral activity in vitro. Acta pathol. microbiol. scand., B, Microbiol. *84:* 280–284 (1976).

DeChatelet, L.R.; Cooper, M.R.; McCall, C.E.: Stimulation of the hexose monophosphate shunt in human neutrophils by ascorbic acid: mechanisms of action. Antimicrob. Agents Chemother. *1:* 12–16 (1972).

DeChatelet, L.R.; McCall, C.E.; Cooper, M.R.; Shirley, P.S.: Ascorbic acid levels in phagocytic cells. Proc. Soc. exp. Biol. Med. *145:* 1170–1173 (1974).

Delafuente, J.C.; Dlesk, A.; Panush, R.S.: Cellular immunity; in Panush, Principles of rheumatic diseases, pp. 99–118 (Wiley, New York 1981).

Delafuente, J.C.; Panush, R.S.: Enhancement of in vitro lymphocyte responses by ascorbic acid (Abstract). J. Allergy clin. Immunol. *61:* 132 (1978).

Delafuente, J.C.; Panush, R.S.: Modulation of certain immunologic responses by vitamin C. II. Enhancement of concanavalin A-stimulated lymphocyte responses. Int. J. Vitam. Nutr. Res. *50:* 44–51 (1980).

Dennert, G.; Lotan, R.: Effects of retinoic acid on the immune system: stimulation of T killer cell induction. Eur. J. Immunol. *8:* 23–29 (1978).

Destro, R.L.; Sharma, V.: An appraisal of vitamin C in adjunct therapy of bacterial and 'viral' meningitis. Clin. Pediat. *16:* 936–939 (1977).

Donnall, E.; Baker, J.A.; Ferrebee, J.W.: The effect of methotrexate on the production of antibodies against attenuated distemper virus in the dog. J. Immun. *90:* 324–328 (1963).

Dykes, M.H.M.; Meier, P.: Ascorbic acid and the common cold. Evaluation of its efficacy and toxicity. J. Am. med. Ass. *231:* 1073–1079 (1975).

Editorial: Vitamin C and the common cold. Br. med. J. *i:* 606 (1976).

Editorial: Ascorbic acid: immunologic effects and hazards. Lancet *i:* 308 (1979).

Falchuk, K.R.; Walker, W.A.; Perrotto, J.L.; Isselbacher, K.J.: Effect of vitamin A on the systemic and local antibody response to intragastrically administered bovine serum albumin. Infect. Immunity *17:* 361–365 (1977).

Felix, E.L.; Loyd, B.; Cohen, M.H.: Inhibition of the growth and development of a transplantable murine melanoma by vitamin A. Science *189:* 886–889 (1975).

Fidanza, A.; Audisio, M.; Mastroiacovo, P.: Vitamin C and cholesterol. Int. J. Vitam. Nutr. Res. *23:* suppl., pp. 173–186 (1982).

Fisher, B.; Axelrod, A.E.; Fisher, E.R.; Lee, S.H.; Calvanese, N.: The favorable effect of pyridoxine deficiency on skin homograft survival. Surgery, St. Louis *44:* 149–167 (1958).

Fortner, B.R., Jr.; Danziger, R.E.; Rabinowitz, P.S.; Nelson, H.S.: The effect of ascorbic acid on cutaneous and nasal response to histamine and allergen. J. Allergy clin. Immunol. *69:* 484–488 (1982).

Fraser, R.C.; Pavlovic, S.; Kurahara, C.G.; Murata, A.; Peterson, N.S.; Taylor, K.B.; Feigen, G.A.: The effect of variations in vitamin C intake on the cellular immune response of guinea pigs. Am. J. clin. Nutr. *33:* 839–847 (1980).

Friedenberg, W.R.; Marx, J.J.; Hansen, R.L.; Haselby, R.C.: Hyperimmunoglobulin E syndrome: response to transfer factor and ascorbic acid therapy. Clin. Immunol. Immunopathol. *12:* 132–142 (1979).

Friedman, R.M.; Buckler, C.E.; Baron, S.: The effect of aminomethylpteroylglutamic acid on the development of skin hypersensitivity and on antibody formation in guinea pigs. J. exp. Med. *114:* 173–183 (1961).

Fulghum, D.D.: Ascorbic acid revisited. Archs Derm. *113:* 91–92 (1977).

Gallin, J.I.; Elin, R.J.; Hubert, R.T.; Fauci, A.S.; Kaliner, M.A.; Wolff, S.M.: Efficacy of ascorbic acid in Chediak-Higashi syndrome (CHS): studies in humans and mice. Blood *53:* 226–234 (1979).

Gallin, J.I.; Sandler, J.A.; Clyman, R.I.; Manganiello, V.C.; Vaughan, M.: Agents that increase cyclic AMP inhibit accumulation of cGMP and depress human monocyte locomotion. J. Immun. *120:* 492–496 (1978).

Ganguly, R.; Durieux, M.F.; Waldman, R.H.: Macrophage function in vitamin C-deficient guinea pigs. Am. J. clin. Nutr. *29:* 762–765 (1976).

Gatner, E.M.; Anderson, R.: An in vitro assessment of cellular and humoral immune function in motility by ascorbate, levamisole, metoprolol and propranolol. Clin. exp. Immunol. *40:* 327–336 (1980).

Ginter, E.; Bobek, P.; Babala, J.; Jakobovsky, J.; Zaviacic, M.; Lojda, Z.: Vitamin C in atherosclerosis. Int. J. Vitam. Nutr. Res. *19:* suppl., pp. 55–90 (1979).

Ginter, E.; Bobek, P.; Kubec, F.; Vozar, J.; Urbanova, D.: Vitamin C in the control of hypercholesterolemia in man. Int. J. Vitam. Nutr. Res. *23:* suppl., pp. 153–172 (1982).

Goetzl, E.J.: Defective responsiveness to ascorbic acid of neutrophil random and chemotactic migration in Felty's syndrome and systemic lupus erythematosus. Ann. rheum. Dis. *35:* 510–515 (1976).

Goetzl, E.J.; Wasserman, S.I.; Austen, K.F.: Modulation of the eosinophil chemotaxis response in immediate hypersensitivity; in Brent, Hoborow, Progression in immunology II, vol. 4, pp. 41–50 (North-Holland, Amsterdam 1974a).

Goetzl, E.J.; Wasserman, S.I.; Gigli, I.; Austen, K.F.: Enhancement of random migration of chemotactic responses of human leukocytes by ascorbic acid. J. clin. Invest. *53:* 813 (1974b).

Gonzales, R.G.; Garcia, M.; Perez, S.H.; Vega, A.R. de la: Comparative study of ascorbic acid and disodium cromoglycate in some models of experimental anaphylaxis. Allerg. Immunopathol. *7:* 211–216 (1979).

Gross, R.L.; Newberne, P.M.: Role of nutrition and immunologic function. Physiol. Rev. *60:* 188–302 (1980).

Hahn, G.S.; O'Connor, R.: Sodium ascorbate stimulation of phytohemagglutinin induced T lymphocyte mitogenesis (Abstract). Clin. Res. *26:* 184A (1978).

Hall, M.G.; Darling, R.C.; Taylor, F.H.L.: The vitamin C requirement in rheumatoid arthritis. Ann. intern. Med. *13:* 415–423 (1939).

Hanck, A.: Vitamin C. New clinical applications in immunology, lipid metabolism and cancer. Int. J. Vitam. Nutr. Res. *23:* suppl., pp. 7–294 (1982).

Hanck, A.; Ritzel, G.: Vitamin C. Recent advances and aspects in virus disease, cancer and in lipid metabolism. Int. J. Vitam. Nutr. Res. *19:* suppl., pp. 7–212 (1979).

Hanck, A.; Weiser, H.: The influence of vitamin C on lipid metabolism in man and animals. Int. J. Vitam. Nutr. Res. *19:* suppl., pp. 83–94 (1979).

Heine, H.; Norden, C.: Vitamin C therapy in hyperlipoproteinemia. Int. J. Vitam. Nutr. Res. *19:* suppl., pp. 45–54 (1979).

Heinzerling, R.H.; Tengerdy, R.P.; Wick, L.L.; Leuker, D.C.: Vitamin E protects mice against *Diplococcus pneumoniae* type I infection. Infect. Immunity *10:* 1292–1295 (1974).

Hitchcock, M.: Effect of variation in endogenous levels of ascorbic acid on the in vitro immunological release of histamine and slow reacting substance of anaphylaxis from actively sensitized guinea-pig lung fragments. Br. J. Pharmacol. *11:* 539–543 (1980).

Hodges, R.E.; Bean, W.B.; Ohlson, M.A.; Bleiler, R.E.: Factors effecting human antibody responses. V. Combined deficiencies of pantothenic acid and pyridoxine. Am. J. clin. Nutr. *11:* 187–199 (1962).

Hornig, D.: Metabolism of ascorbic acid. Wld Rev. Nutr. Diet., vol. 23, pp. 225–258 (Karger, Basel 1975).

Hsu, C.-K.: Vitamin C and immune responses in rhesus monkeys (Abstract). Fed. Proc. *36:* 1177 (1977).

Jukes, T.H.: Further comments on the ascorbic acid requirement. Proc. natn. Acad. Sci. USA *72:* 4151–4152 (1975).

Jurin, M.; Tannock, I.F.: Influence of vitamin A on immunological response. Immunology *23:* 283–287 (1972).

Kaiser, A.D.: Rheumatic infection. Is vitamin C deficiency a factor? N.Y. St. med. J. *38:* 868–873 (1938).

Kalden, J.R.; Guthy, E.A.: Prolonged skin allograft survival in vitamin C-deficient guinea pigs. Eur. surg. Res. *4:* 114–119 (1972).

Karlowski, T.; Chalmers, T.C.; Frenkel, L.D.; Kapikian, A.Z.; Lewis, J.M.; Lynch, J.M.: Ascorbic acid and the common cold. A prophylactic and therapeutic trial. J. Am. med. Ass. *231:* 1038–1042 (1975).

Karpinska, T.; Kawecki, Z.; Kandefer-Szerszen, M.: The influence of ultraviolet irradiation, *L*-ascorbic acid and calcium chloride on the induction of interferon in human embryo fibroblasts. Arch. Immunol. Ther. Exp. (Warsz.) *30:* 33–37 (1982).

Kay, N.E.; Holloway, D.E.; Hutton, S.W.; Bona, N.D.; Duane, W.C.: Human T-cell function in experimental ascorbic acid deficiency and spontaneous scurvy. Am. J. clin. Nutr. *36:* 127–130 (1982).

King, C.G.; Burns, J.J.: Second conference on vitamin C. Ann. N.Y. Acad. Sci. *258:* 2–552 (1975).

Knodell, R.G.; Tate, M.A.; Arl, B.F.; Wilson, J.W.: Vitamin C prophylaxis for post-transfusion hepatitis: lack of effect in a controlled trial. Am. J. clin. Nutr. *34:* 20–23 (1981).

Kordansky, D.W.; Rosenthal, R.R.; Norman, P.S.: The effect of vitamin C on antigen-induced bronchospasm. J. Allergy clin. Immunol. *63:* 61–64 (1979).

Kraut, E.H.; Metz, E.N.; Sagane, A.L.: In vitro effects of ascorbate on white cell metabolism and the chemiluminescence response. J. reticuloendoth. Soc. *27:* 359–366 (1980).

Krishnan, S.; Bhuyan, U.N.; Talwar, G.P.; Ramalingaswami, V.: Effect of vitamin A and protein-calorie undernutrition on immune responses. Immunology *27:* 383–392 (1974).

Kumar, M.; Axelrod, A.E.: Circulating antibody formation in scorbutic guinea-pigs. J. Nutr. *98:* 41–44 (1969).

Manzella, J.P.; Roberts, N.J.: Human macrophage and lymphocyte responses to mitogen stimulation after exposure to influenza virus, ascorbic acid, and hyperthermia. J. Immun. *123:* 1940–1944 (1979).

Massell, B.F.; Warren, J.E.; Patterson, P.R.; Lehmus, H.J.: Antirheumatic activity of ascorbic acid in large doses. Preliminary observations on seven patients with rheumatic fever. New Engl. J. Med. *242:* 614–615 (1950).

McCorkle, F.; Taylor, R.; Stinson, R.; Day, E.J.; Glick, B.: The effects of a megalevel of vitamin C on the immune response of the chicken. Poult. Sci. *59:* 1324–1327 (1980).

Medawar, P.B.; Hunt, R.: Anti-cancer action of retinoids. Immunology *42:* 349–353 (1981).

Mueller, P.S.; Kies, M.W.: Suppression of tuberculin reaction in the scorbutic guinea pigs. Nature, Lond. *195:* 813 (1962).

Mueller, P.S.; Kies, M.W.; Alvord, E.C.; Shaw, C.-M.: Prevention of experimental allergic encephalomyelitis (EA) by vitamin C deprivation. J. exp. Med. *115:* 329–338 (1962).

Munster, A.M.; Loadholdt, C.B.; Barnes, M.A.: The effect of antibiotics on cell-mediated immunity. Surgery, St. Louis *81:* 692–695 (1977).

Murata, A.; Morishige, F.; Yamaguchi, H.: Prolongation of survival times of terminal cancer patients by administration of large doses of ascorbate. Int. J. Vitam. Nutr. Res. *23:* suppl., pp. 113–114 (1982).

Nauss, K.M.; Mark, D.A.; Suskind, R.M.: The effect of vitamin A deficiency on the in vitro cellular immune response of rats. J. Nutr. *109:* 1815–1823 (1979).

Nungester, W.J.; Ames, A.M.: The relationship between ascorbic acid and phagocytic activity. J. infect. Dis. *83:* 50–54 (1948).

Ogilvy, C.S.; DuBois, A.B.; Douglas, J.S.: Effects of ascorbic acid and indomethacin on the airways of healthy subjects with and without inferred bronchoconstriction. J. Allergy clin. Immunol. *67:* 363–369 (1981).

Olson, G.E.; Polk, H.C.: In vitro effect of ascorbic acid on corticosteroid-caused neutrophil dysfunction. J. surg. Res. *22:* 109–112 (1977).

Panush, R.S.; Delafuente, J.C.: Modulation of certain immunologic responses by vitamin C. Int. J. Vitam. Nutr. Res. *19:* suppl., pp. 179–199 (1979).

Panush, R.S.; Delafuente, J.C.; Katz, P.; Johnson, J.: Modulation of certain immunologic response by vitamin C. III. Potentiation of in vitro and in vivo lymphocyte responses. Int. J. Vitam. Nutr. Res. *23:* 35–47 (1982).

Panush, R.S.; Katz, P.; Powell, G.; Somberg, L.: Immunopharmacologic effects of vitamin C. IV. Perturbation of mononuclear cell cyclic nucleotides. Int. J. Vitam. Nutr. Res. *53:* 61–67 (1983).

Patrone, F.; Dallegri, F.; Bonvini, E.; Minervini, F.; Sacchetti, C.: Effects of ascorbic acid on neutrophil function. Studies of normal and chronic granulomatous disease neutrophils. Acta vitaminol. enzymol. *4:* 163–166 (1982a).

Patrone, F.; Dallegri, F.; Minervini, F.; Sacchetti, C.: Disorders of neutrophil function in children with recurrent pyogenic infections. Med. Microbiol. Immunol. *171:* 113–122 (1982b).

Pauling, L.: Vitamin C and the common cold (Freeman, San Francisco 1970).

Pauling, L.: Are recommended daily allowances for vitamin C adequate? Proc. natn. Acad. Sci. USA *71:* 4442–4446 (1974).

Pauling, L.: The role of vitamin C in cancer. Int. J. Vitam. Nutr. Res. *19:* suppl., pp. 207–210 (1979).

Pickett, W.C.; Austen, K.F.; Goetzl, E.J.: Inhibition by non-steroidal anti-inflammatory agents of the ascorbate-induced elevations of platelet cyclic GMP levels. J. cyclic Nucleotide Res. *5:* 197–209 (1979).

Pitt, H.A.; Costrini, A.M.: Vitamin C prophylaxis in marine recruits. J. Am. med. Ass. *241:* 908–911 (1979).

Prasad, J.S.: Effect of vitamin E supplementation on leukocyte function. Am. J. clin. Nutr. *33:* 606–608 (1980).

Prinz, W.; Bloch, J.; Gilich, G.; Mitchell, G.: A systemic study of the effect of vitamin C supplementation on the humoral immune response in ascorbate-dependent mammals. I. The antibody response to sheep red blood cells (a T-dependent antigen) in guinea pigs. Int. J. Vitam. Nutr. Res. *50:* 294–300 (1980).

Prinz, W.; Bortz, R.; Hersch, M.; Gilich, G.: Vitamin 'C' and the humoral immune response. Int. J. Vitam. Nutr. Res. *19:* suppl., pp. 25–34 (1979).

Ramirez, I.; Riebie, E.; Waag, Y.-M.; Van Eys, J.: Effect of ascorbic acid in vitro on lymphocyte reactivity to mitogens. J. Nutr. *110:* 2207–2215 (1980).

Rebora, A.; Crovato, F.; Dallegri, F.; Patrone, F.: Repeated staphylococcal pyoderma in two siblings with defective neutrophil bacterial killing. Dermatologica *160:* 106–112 (1980a).

Rebora, A.; Dallegri, F.; Patrone, F.: Neutrophil dysfunction and repeated infections: influence of levamisole and ascorbic acid. Br. J. Derm. *1021:* 49–56 (1980b).

Repine, J.E.; Rao, G.; Beall, G.D.; White, J.G.: Inhibition of human neutrophil oxidative

metabolism and degranulation in vitro by nitroblue tetrazolium and vitamin E. Am.
J. Path. *90:* 659–674 (1978).

Rhodes, J.; Oliver, S.: Retinoids as regulators of macrophage function. Immunology *40:* 467–472 (1980).

Rinehart, J.F.: Studies relating vitamin C deficiency to rheumatic fever and rheumatoid arthritis: experimental, clinical and general considerations. II. Rheumatoid arthritis. Ann. intern. Med. *9:* 671–688 (1935).

Rinehart, J.F.; Greenberg, L.D.; Baker, F.; Mettler, S.R.; Brockman, F.; Choy, F.: Metabolism of vitamin C in rheumatoid arthritis. Archs intern. Med. *61:* 552–561 (1938).

Saitoh, H.; Komiyama, A.; Norose, N.; Morosawa, H.; Akabane, T.: Development of the accelerated phase during ascorbic acid therapy in Chediak-Higashi syndrome and efficacy of colchicine on its management. Br. J. Haemat. *48:* 79–84 (1981).

Sandler, J.A.; Gallin, J.I.; Vaughan, M.: Effects of carbamylcholine and ascorbic acid on leukocyte cyclic GMP and chemotaxis. J. Cell Biol. *67:* 480–484 (1975a).

Sandler, J.A.; Smith, T.K.; Manganiello, V.C.; Kirkpatrick, C.H.: Stimulation of monocyte cGMP by leukocyte dialysates. J. clin. Invest. *56:* 1271–1279 (1975b).

Schoepflin, G.S.; Goetzl, E.J.; Austen, K.F.: The predominant contribution of platelets to baseline and ascorbate-stimulated increments in cyclic GMP in human mononuclear cell preparations. Cell. Immunol. *35:* 330–339 (1978).

Schorah, C.J.; Newill, A.; Scott, D.L.; Morgan, D.B.: Clinical effects of vitamin C in elderly patients with low blood vitamin C levels. Lancet *i:* 403–405 (1979).

Schorah, C.J.; Tormey, W.P.; Brooks, G.H.; Robertshaw, A.M.; Young, G.A.; Talukder, R.; Kelly, J.F.: The effect of vitamin C supplements on body weight, serum proteins, and general health of an elderly population. Am. J. clin. Nutr. *34:* 871–876 (1981).

Sharma, S.C.; Wilson, C.W.: The cellular interaction of ascorbic acid with histamine, cyclic nucleotides and prostaglandins in the immediate hypersensitivity reaction. Int. J. Vitam. Nutr. Res. *50:* 163–170 (1980).

Shilotri, P.G.: Glycolytic, hexose monophosphate shunt and bactericidal activities of leukocytes in ascorbic acid deficient guinea pigs. J. Nutr. *107:* 1507–1512 (1977).

Shilotri, P.G.; Bhat, K.S.: Effect of mega doses of vitamin C on bactericidal activity of leukocytes. Am. J. clin. Nutr. *30:* 1077–1081 (1977).

Siegel, B.V.: Enhanced interferon response to murine leukemia virus by ascorbic acid. Infect. Immunity *10:* 409–410 (1974).

Siegel, B.V.: Enhancement of interferon productivity by poly (RI) poly (RC) in mouse cell cultures by ascorbic acid. Nature, Lond. *244:* 531–532 (1975).

Siegel, B.V.; Morton, J.I.: Vitamin C and the immune response. Experientia *33:* 393–395 (1977).

Smith, W.B.; Shohet, S.B.; Zagajeski, E.; Lubin, B.H.: Alteration in human granulocyte function after in vitro incubation with *L*-ascorbic acid. Ann. N.Y. Acad. Sci. *258:* 329–338 (1975).

Stankova, L.; Gerhardt, N.B.; Nagel, L.; Bigley, R.H.: Ascorbate and phagocyte dysfunction. Infect. Immunity *12:* 252–256 (1975).

Stein, H.B.; Hasan, A.; Fox, I.H.: Ascorbic acid-induced uricosuria. Ann. intern. Med. *84:* 385–388 (1976).

Tanaka, J.; Fujiwara, H.; Torisu, M.: Vitamin E and immune response. I. Enhancement of helper T cell activity by dietary supplementation of vitamin E in mice. Immunology *38:* 727–734 (1979).

Tengerdy, R.P.: Effect of vitamin E on immune responses; in Machlin, Vitamin E, pp. 429–444 (Dekker, New York 1980).

Tengerdy, R.P.; Heinzerling, R.H.; Brown, G.L.; Mathias, M.M.: Enhancement of the humoral immune response by vitamin E. Int. Archs Allergy appl. Immun. *44:* 221–232 (1973).

Tengerdy, R.P.; Heinzerling, R.H.; Nockel, C.F.: Effect of vitamin E on the immune response of hypoxic and normal chickens. Infect. Immunity *5:* 987–989 (1972).

Thomas, W.R.; Holt, P.G.: Vitamin C and immunity: an assessment of the evidence. Clin. exp. Immunol. *32:* 370–379 (1978).

Thurman, G.B.; Goldstein, A.L.: Suppression of immunological responsivity in guinea pigs by ascorbic acid depletion (Abstract). Fed. Proc. *38:* 1173 (1979).

Tolbert, B.M.: Ascorbic acid metabolism and physiological function. Int. J. Vitam. Nutr. Res. *19:* suppl., pp. 127–142 (1979).

Trakatellis, A.C.; Stinebring, W.R.; Axelrod, A.E.: Studies on systemic reactivity to purified protein derivative and endotoxin. I. Systemic reactivity to PPD in pyridoxine-deficient guinea pigs. J. Immun. *91:* 39–45 (1963).

Tsien, W.-H.; Sampson, M.; Sheppard, H.: Effects of ascorbic acid on ^{3}H-thymidine incorporation by isolated mouse thymocytes. Immunopharmacology *2:* 117–130 (1980).

Vallance, S.: Relationships between ascorbic acid and serum proteins of the immune system. Br. med. J. *2:* 437–438 (1977).

Vilter, R.W.: Nutritional aspects of ascorbic acid: uses and abuses. West. J. Med. *133:* 485–492 (1980).

Waldman, R.H.; Ganguly, R.; Gallagher, E.; Durieux, M.F.: Effects of orange juice on viral infections (Abstract). Am. J. Epidemiol. *102:* 466 (1975).

Weening, R.S.; Schoorel, E.P.; Roos, D.; van Schaik, M.L.; Voetman, A.A.; Bot, A.A.; Batenburg-Plenter, A.M.; Willems, C.; Zeijlemaker, W.P.; Astaldi, A.: Effect of ascorbate on abnormal neutrophil, platelet and lymphocyte function in a patient with Chediak-Higashi syndrome. Blood *57:* 856–865 (1981).

Wilson, C.W.: Clinical pharmacological aspects of ascorbic acid. Ann. N.Y. Acad. Sci. *258:* 355–376 (1975).

Wilson, C.W.M.: The controlling effect of ascorbic acid on the actions of anti-obesity and hypocholesterolemic drugs. Int. J. Vitam. Nutr. Res. *19:* suppl., pp. 71–82 (1979).

Zuskin, E.; Lewis, A.J.; Bouhuys, A.: Inhibition of histamine-induced airway constriction by ascorbic acid. J. Allergy clin. Immunol. *51:* 218–226 (1973).

Zweiman, B.; Besdine, R.W.; Hildreth, E.A.: The effect of the scorbutic state on tuberculin hypersensitivity in the guinea pig. II. In vitro mitotic response of lymphocytes. J. Immun. *96:* 672–675 (1966a).

Zweiman, B.; Schoenwetter, W.F.; Hildreth, E.A.: The effect of the scorbutic state of tuberculin hypersensitivity in the guinea pig. I. Passive transfer of tuberculin hypersensitivity. J. Immun. *96:* 296–300 (1966b).

Dr. Richard S. Panush, Division of Clinical Immunology,
Department of Medicine, Box J-277, University of Florida,
Gainesville, FL 32610 (USA)

Wld Rev. Nutr. Diet., vol. 45, pp. 133–166 (Karger, Basel 1985)

Vitamin E and Blood[1]

Ching K. Chow

Department of Nutrition and Food Science, University of Kentucky,
Lexington, Ky., USA

Contents

Introduction

Vitamin E is the term suggested for all toco and tocotrienol derivatives exhibiting qualitatively the biological activity of α-tocopherol. All eight naturally occurring compounds known in the tocopherol series are derivatives of 6-chromanol (fig. 1). This series is made up of 4 compounds with a toco structure which bears a saturated isoprenoid C_{16}-side chain and 4 compounds with a tocotrienol structure bearing 3 double bonds in the isoprenoid-side chain.

[1] Literature reviewed up to May, 1983.

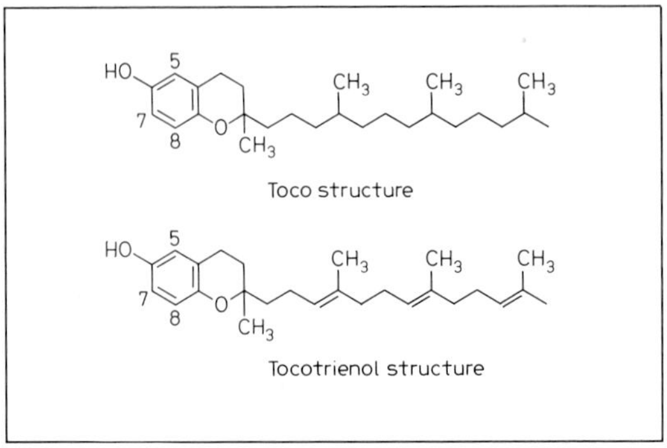

Fig. 1. Structural formula of tocopherols.

Tocopherols are pale yellow or yellow-tan viscous liquids at room temperature. At high purity they are relatively odorless and colorless. They are lipid-soluble and, hence, fully soluble in lipoidal solvents and insoluble in water. The tocopherols can be oxidized to form quinones, tocoreds, dimers, trimers and other products. The process is accelerated in the presence of light, heat, alkali and certain divalent metals such as iron and copper. In the absence of oxygen they are relatively stable to heat, light and alkali. Esters that acylate the free phenolic hydroxy group increase the oxidative stability of the compounds [*Bauernfeind and Cort,* 1974].

The tocopherols occur mainly in the free alcohol form in a variety of plant life, such as nuts, seeds, oils, fruits and vegetables [*Parrish,* 1980]. In general, seeds, grains and vegetable oils are superior sources to fruit, vegetable or animal tissues and products. The distribution pattern of tocopherols in plant life is modified by species, variety, stage of maturation, climate conditions, harvesting, and processing procedures [*Bauernfeind,* 1977].

Vitamin E was discovered approximately 60 years ago as a fat-soluble dietary factor effective in preventing fetal death and resorption in the rat. Since then, considerable research on various aspects of the vitamin has been conducted and volumes of reports on the subject have been published. While the essentiality of vitamin E in animal nutrition has been well-recognized, its precise role in human nutrition has not yet been firmly established. However, despite of a lack of complete understanding on the mode of action and of a demonstrable bona fide human deficiency syndrome, interest in cellular metabolism and the application of vitamin E in clinical and preventive medicine remains very strong.

Table I. Vitamin E activity of tocopherols (bioassay method)

Structure	Resorption-gestation, %	Hemolysis, %
α-Tocopherol	100	100
β-Tocopherol	25–40	15–27
γ-Tocopherol	8–19	3–20
δ-Tocopherol	0.1–1	0.3–2
α-Tocotrienol	21	17–25
β-Tocotrienol	4	1–5
γ-Tocotrienol	–	–
δ-Tocotrienol	–	–

Blood occupies a unique and important place in our current understanding of vitamin E. Association of low vitamin E levels in plasma (or serum) with erythrocyte hemolysis is the best known manifestation of dietary vitamin E deprivation in experimental animals and humans. Due to easy accessibility and fewer organelle complications, more information regarding the functions of vitamin E has been obtained from the blood than any other organ or organ system of humans.

All blood cells originate from undifferentiated mesenchymal cells. From there, stem cells differentiate and ultimately appear in the circulating blood as red cells, platelets and various types of white cells. Cells of each type possess distinct and specific physiological characteristics and functions of their own. The broad influence of vitamin E on biological processes indicates that the vitamin may play a role in modulating the functions and life span of various types of blood cells. This review will focus on the hematological aspects of vitamin E from the substantial amount of information accumulated during the past decade.

Biological Activity

The main criteria used in the biological evaluation of the tocopherols have been the resorption-gestation assay, the muscular dystrophy score, the erythrocyte hemolysis test, and determination of tocopherol levels in plasma and tissues (mainly liver). A summary of biopotency of the various tocopherols using resorption gestation assay and the hemolysis test is shown in table I. The usual order of biological activity of tocopherols orally

Table II. Tocopherol content of major fats and oils used in food products in the USA[1]

	% of total fats and oils	Total tocopherol mg/100 g oil or fat	Individual tocopherol, % of total[2]							
			α-T	α-T-3	β-T	β-T-3	γ-T	γ-T-3	δ-T	δ-T-3
Animal fats										
Lard	13.3	0.6–1.3	>90	<5			<5			
Butter	8.2	1.0–5.0	>90				<10			
Tallow	4.9	1.5–2.4	>90				<10			
Vegetable oils										
Soybean	53.4	56–160	4–18				58–69		24–37	
Cotton seed	8.9	30–81	51–67				33–49			
Corn	3.8	53–162	11–24				76–89			
Coconut	3.1	1–4	14–67	<14		<3		<53	<17	
Peanut	1.4	20–32	48–61				39–52			
Palm	1.0	33–73	28–50	16–19		4		34–39	<9	
Palm kernel	0.7	0								
Safflower	0.7	25–49	80–94				6–20			
Olive	0.6	5–15	65–85				15–35			
Per capita, lb	53.4									

[1] *Chow* [1975 b].
[2] α-T, β-T, γ-T, and δ-T are α-, β-, γ-, and δ-tocopherol, respectively; α-T-3, β-T-3, γ-T-3, and δ-T-3 are the corresponding tocotrienols.

is $\alpha > \gamma > \beta > \delta$. While the term vitamin E refers to at least 8 tocopherol structures possessing vitamin E activity, α-tocopherol predominates in many species and is significantly more potent than any other naturally occurring tocopherols known. Therefore, a determination of α-tocopherol content is usually a good approximation of the total vitamin E activity of the food or feed source.

Since α-tocopherol has the highest biological activity of the naturally occurring forms of vitamin E, vitamin E status in human population is usually assayed in terms of the α-tocopherol level in blood. However, based on the content of individual tocopherols present in the major edible fats and analysis of representative American meals, γ-tocopherol is estimated to constitute over 50% of total tocopherol intake in the American population, while α- and δ-tocopherol account for approximately 20% each (table II).

Although non-α-tocopherols have been shown to be biologically less active as compared with the α-form, muscular dystrophy in chicks has been found to be prevented by similar concentrations of β-, γ- and δ-tocopherol in the plasma [*Scott and Desai,* 1964]. Therefore, the contribution of non-α-tocopherols toward the human need of vitamin E may be very important, especially in the absence of the α-form.

Several procedures/methods have been suggested and/or employed for the analysis of vitamin E. The conventional method for resolving different isomers of tocopherols usually involves the extraction of a sample to be analyzed with an appropriate organic solvent, separation of tocopherols from the bulk of lipid components by saponification, and low-temperature crystallization. This is followed by thin-layer chromatographic separation and quantitation spectrophotometrically with iron reagents (α,α-dipyridine or bathophenanthroline) or by fluorometric or gas-liquid chromatographic technique [*Barnes and Taylor,* 1981].

The analytical procedure for vitamin E using iron reagents lacks specificity because other substances are often present that will reduce the resultant chromogenic complex. Furthermore, the individual tocopherols react at varying rates to produce the reduced compound which is measured spectrophotometrically. While the specificity can be partially improved by elaborate sample preparation, such as column or thin-layer chromatography to remove reducing artifacts, the cumbersome techniques usually lead to poor reproducibility and/or recoveries. During the past few years, the development of high-pressure liquid chromatographic techniques has made it possible to accurately measure extremely small quantities (as low as 1 ng) of individual tocopherols in biological samples, as well as in foods and feeds [*Tangney* et al., 1981; *Hatam and Kayden,* 1979; *Bieri* et al., 1979; *Soderhjelm and Anderson,* 1978].

Nutritional Status

Vitamin E nutritional status in animals has been assessed by tissue analysis of tocopherols and, in certain circumstances, by evaluating one or more tests of vitamin E function. Analysis of plasma or serum levels of vitamin E has been the most frequent method used when assessing the current status of vitamin E, especially in human subjects. In experimental animals, analysis of tocopherol levels in liver, adipose tissue or other organs is valuable in estimating total body reserves of vitamin E. Tests of vitamin

Table III. Pathology of vitamin E deficiency

Condition	Animal	Tissue affected
Reproductive failure		
Embryonic degeneration	female: rat, hen, turkey, ewe	vascular system of embryo
Sterility	male: rat, guinea pig, hamster, dog, cock	male gonads
Liver, blood, brain, capillaries		
Liver necrosis	rat, pig	liver
Erythrocyte destruction	rat, chick	blood (erythrocyte hemolysis)
Blood protein loss	chick, turkey	serum albumin
Encephalomalacia	chick	cerebellum (Purkinje cells)
Exudative diathesis	chick, turkey	capillary walls
Kidney degeneration	rat, monkey, mink	tabular epithelium
Steatitis	mink, pig, chick	depot fat
Nutritional myopathies		
Nutritional muscular dystrophy	rabbit, guinea pig, monkey, duck, rat, chick, turkey	skeletal muscle
Stiff lamb	lamb, kid	skeletal muscle
White muscle disease	calf, sheep, mouse, mink	skeletal and heart muscle
Myopathy of gizzard and heart	turkey poult	gizzard, heart, skeletal muscle

E function include the measurement of erythrocyte hemolysis, either spontaneously or in the presence of such chemicals as hydrogen peroxide or dialuric acid [*Mino* et al., 1981; *Bieri and Poukka,* 1970], and measurements of protection of vitamin E against such specific deficient symptoms as muscular dystrophy in rabbits, chicks or rats, and encephalomalacia in chicks [*Scott,* 1969] (table III).

Although several methods have been developed for determining the nutritional status of vitamin E, an accurate method for the evaluation of human vitamin E status has not yet been established. Blood levels of tocopherols widely utilized by many investigators may not accurately reflect either level of intake or tissue storage. While a high value of erythrocyte hemolysis in vitro appears to serve as an indicator of high probability of inadequacy, low hemolysis values do not necessarily indicate adequacy of tissue storage [*Bieri and Poukka,* 1970]. Furthermore, the in vitro hemolysis and other tests are known to be influenced by factors other than vitamin E intake [*Stocks and Dormandy,* 1971; *Melhorn* et al., 1971]. Recently, measurement of tocopherol levels in needle aspiration biopsies of adipose tissue has been suggested to be a good index of vitamin E status in humans [*Hatam and Kayden,* 1982]. However, the procedure may not be easily applicable by every investigator due to the nature of sampling technique.

The characteristic enzyme pattern of plasma is often altered when enzymes leak out of diseased cells. Detection of an unusual enzyme pattern or origin-specific enzymes has been widely used as a sensitive index of tissue damage or disease. The activities of a number of enzymes including several lysosomal enzymes have been found to be altered in plasma of vitamin E-deficient animals [*Zalkin* et al., 1962]. The activity of pyruvate kinase, for example, has been shown to be markedly increased in the plasma of vitamin E-deficient rats and has been shown to be a sensitive index of myopathy resulting from vitamin E deficiency [*Chow,* 1975a; *Gabriel* et al., 1980]. However, alterations in activities of specific enzymes in plasma resulting from changing nutritional status of vitamin E in humans remain to be demonstrated.

In the USA, a recommended daily dietary allowance (RDA) for vitamin E was first set in 1965 [Recommended Dietary Allowance, 1965]. However, the initial values could not be met by most US diets adequate in all other nutrients. It has been shown that consumption of typical balanced diets of about 2,500 kcal contained from 6.4 to 9.0 mg (9.5–13.4 IU) of α-tocopherol daily [*Bieri and Evarts,* 1973; *Bieri and Farrell,* 1976]. Non-α-tocopherols have been estimated to contribute only approximately 20% of total vitamin E activity even though at least 3 times as much as the α-form have been consumed [*Bieri and Evarts,* 1973; *Chow,* 1975b]. Therefore, the RDA for vitamin E was reduced, and age and sex were considered in subsequent editions of RDA [Recommended Dietary Allowance, 1974; 1980]. Based on the dietary analysis, a recommended range of 12–15 IU of total vitamin E activity should be present in the average diet of adults consuming 2,000–3,000 kcal.

Dietary variables such as polyunsaturated fatty acid content may alter the degree of vitamin E adequacy. Efforts have been made to determine the adequacy ratio between vitamin E and polyunsaturated fatty acids [*Harris and Embree,* 1963; *Horwitt* et al., 1972; *Witting and Lee,* 1975]. However, a fixed vitamin E to polyunsaturated fatty acid ratio does not apply in a variety of experimental conditions, and factors other than polyunsaturated fatty acid content also effect the vitamin E status [*Jager,* 1972].

Absorption and Turnover

Vitamin E has been shown to be absorbed from the intestine as a free tocopherol, in the same manner as fat absorption [*Bieri and Farrell,* 1976]. Esterified tocopherols are hydrolyzed in the intestine, and only the free forms appear in lymph. The overall absorption of small doses of tocopherol in man has been reported to be as low as 20–30% when measured in the lymph [*Blomstrand* et al., 1968] and as high as 85% when fecal excretion was determined [*Kelleher and Losowsky,* 1970]. However, it is well established that a marked decrease in absorption efficiency will result as the dose is increased [*Losowsky* et al., 1972]. Contrary to earlier reports which indicated very poor utilization for non-α-tocopherols, γ-tocopherol for example, has been shown to be absorbed at about 85% that of the α-form [*Gloor* et al., 1966]. Also, γ-tocopherol has been shown to be taken up by the tissues almost as efficiently as the α-form [*Gloor* et al., 1966; *Peake and Bieri,* 1971].

Studies of the kinetics of α-tocopherol depletion and repletion in various tissues of rats indicate that after the rapid growth phase of young animals, relatively constant concentrations are maintained in each tissue. The concentrations are generally related to the log of dietary intake except for adipose tissue where the vitamin E accumulates [*Bieri,* 1972]. When the vitamin was removed from the diet, about one half the α-tocopherol in liver and heart was depleted in 2 weeks, with the remainder disappearing more slowly over a long period [*Bieri,* 1972]. Thus, it appears that there are two pools of α-tocopherol: one labile fraction and one non-labile fraction. The relatively non-labile fraction may represent tocopherol in cell membranes [*Molenaar* et al., 1973].

It has been shown that following administration of high levels (up to 10,000 ppm) of α-tocopherol, tocopherol levels in the tissues analyzed (plasma, platelets, liver, red blood cells, adipose tissue, heart, lung, skeletal muscle and brain) continue to increase for the duration of the supplement up to 20 weeks [*Machlin and Gabriel,* 1982]. Thus, it appears to be difficult to saturate tissue with tocopherol, and also both the level and the duration of supplementation of vitamin E influences the concentration of vitamin E in all tissues. Among the tissues studied, liver has a very rapid rate of accumulation and depletion compared to other tissues [*Machlin and Gabriel,* 1982; *Bieri,* 1972]. The studies suggest that, at least for a short time period, the liver is the major available storage organ for tocopherol.

The tissue levels of a compound usually reflect absorption from the intestine and blood, and its rate of turnover. Since intestinal absorption does not appear to be important in determining the tissue levels of tocopherols, relative turnover rates of various tocopherols may account for the differences in the tissue levels. The relative turnover rates of tocopherols administered intravenously have been shown to be $\beta>\delta>\gamma>\alpha$ in rabbit plasma [*Chow* et al., 1971]. α-Tocopherol has also been shown to be retained better than non-α-tocopherols in rat tissues [*Pearson and McBarnes,* 1970; *Bieri and Evarts,* 1974]. Thus, the striking difference in tocopherol distribution between the estimated intake and the observed values in human blood may in part be attributed to the relatively faster turnover rate of non-α-tocopherols. Since plasma tocopherols are associated with lipoproteins, the turnover rate of tocopherols in plasma may be influenced by a specificity in their binding with lipoproteins.

Metabolism

Because of its antioxidant property, investigation concerning the metabolic fate of vitamin E has been centered initially on the formation and detection of the primary oxidation product of tocopherol, tocopheryl quinone. Evidence for the formation of α-tocopheryl quinone in animal tissues was obtained using radioactive α-tocopherol as the test compound [*Csallany* et al., 1962; *Weber and Wiss,* 1963].

Studies on the metabolism of α-tocopheryl quinone [*Chow* et al., 1967] showed no conversion to α-tocopherol in rat liver. The compound was metabolized partially by reduction to the hydroquinone, conjugation with glucuronic acid, secretions in the bile and elimination in the feces. Another portion of the quinone may have been degraded through a β-oxidation pathway in the kidney to α-tocopheronic acid, followed by conjugation and elimination in the urine (fig. 2). The urinary metabolite was first isolated by *Simon* et al. [1956] after administration of massive dosage of α-tocopherol to rabbits and humans. It is not biologically active in animals and is probably the major end product of tocopherol catabolism under normal conditions.

Studies on the metabolism of α-tocopheryl hydroquinone in rats [*Chow* et al., 1967] showed that it was more rapidly eliminated from the liver than was the corresponding quinone. Conjugates of tocopheronic acid in the urine and of the hydroquinone in the feces were the main excretory forms.

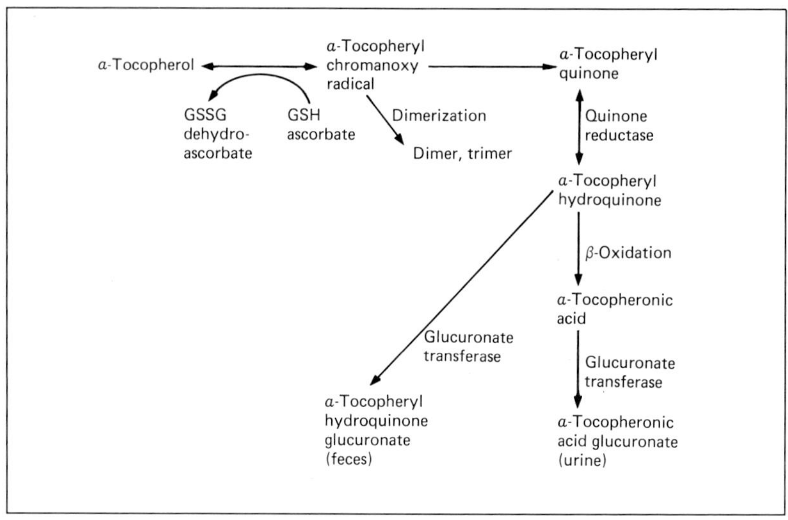

Fig. 2. Metabolic fate of α-tocopherol.

Reduction of the quinone to hydroquinone was found to be closely coupled with conjugation and may thus account for the lack of biological activity of the oxidized form. Since the rate of α-tocopheryl quinone formation from α-tocopherol is rather slow and the efficiency with which it is reduced and conjugated in the liver and kidney is less, it is unlikely that significant amounts of the free hydroquinone exist under normal physiological conditions in animal or human tissues.

In addition to tocopheronic acid, tocopheryl quinone and its hydroquinone, small amounts of dimer and trimer of α-tocopherol have been isolated in animal tissue [*Draper* et al., 1962, 1967; *Csallany and Draper,* 1963]. Little is known concerning the metabolic fate of non-α-tocopherols.

Biological Functions

Although many biochemical abnormalities have been found to be associated with vitamin E deficiency and many investigators are still highly interested in the elucidation of biochemical functions of vitamin E, the exact mechanism as to how tocopherol prevents various metabolic lesions has not yet been proven. Several theories have been proposed, but the nec-

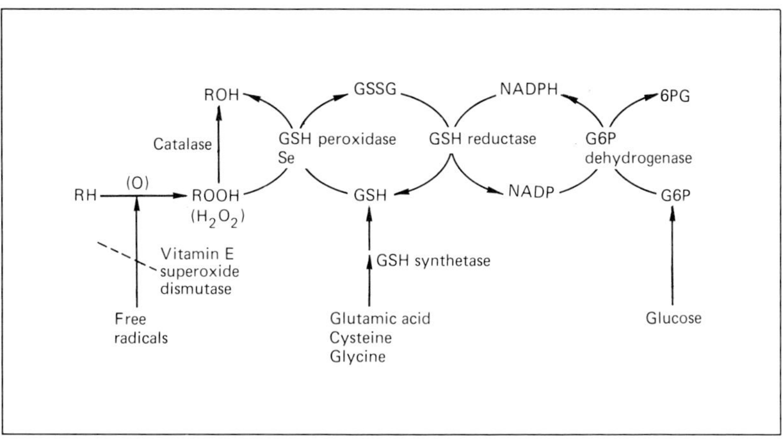

Fig. 3. Important cellular antioxidant defense systems. RH = Polyunsaturated fatty acids; ROOH = fatty acid hydroperoxide; ROH = hydroxy fatty acid; GSH = reduced glutathione; GSSG = oxidized glutathione; SE = selenium, G6P = glucose-6-phosphate; 6PG = 6-phosphogluconate.

essary supportive evidence has yet to be provided [*Green,* 1972; *Tappel,* 1962].

The primary role of vitamin E in preventing lipid peroxidation damage in tissues is accepted by most investigators in the field. This action, however, does not adequately explain some biochemical abnormalities observed in vitamin E deficiency. For example, while there is good evidence that tocopherols may exert their antioxidant function in vivo in adipose tissue, evidence for peroxidative effects resulting from vitamin E deficiency in other tissues is not as certain. The presence of other cellular antioxidant defense systems (fig. 3), especially the glutathione (GSH) peroxidase system, in the parenchyma tissues has been attributed at least partly to the discrepancy [*Chow,* 1979; *Chow and Tappel,* 1972].

Increasing evidence indicates that vitamin E may exert its biological function in relation to cellular membranes. The localization of vitamin E within the membrane as a complex with the polyunsaturated fatty acids of phospholipids [*Lucy,* 1972] may thus inhibit the peroxidation of membrane-bound polyunsaturated fatty acids during electron transport and various other functions [*McCay* et al., 1972]. Vitamin E has been shown to be a part of the inner mitochondrial membrane of duck liver [*Molenaar* et al., 1970] and localized exclusively in membranes of the human red blood cells

[*Chow*, 1975b]. In addition to erythrocyte hemolysis, vitamin E deficiency has been shown to markedly affect the ultrastructure and integrity of mitochondrial, endoplasmic reticular and nuclear membranes of duck liver [*Molenaar* et al., 1970]. Similarly, an adverse effect on the ultrastructure of platelets of vitamin E-deficient rats has been reported [*Arimori and Sumitomo*, 1977]. This membrane function of vitamin E does not conflict with the free radical scavenging mechanism or antioxidant function of vitamin E. However, this theory fails to explain how compounds such as N-methyl-γ-tocopheramine and N,N'-diphenyl-p-phenylenediamine can show the same or similar biological activity [*Draper and Johnson*, 1954]. Furthermore, whether this membrane localization property is unique to α-tocopherol or also shared by other tocopherols in relation to their function remains to be shown.

The observed changes of activities in a number of enzymes in tissues of vitamin E-deficient animals leads to the postulation that vitamin E may have a specific regulatory role in protein synthesis. Among the large number of enzymes that undergo changes in activity during vitamin E deficiency, those receiving the most attention have been muscle creatine kinase and liver xanthione oxidase. Vitamin E deficiency has been shown to markedly increase the turnover rate of creatine kinase [*Olson*, 1974] and the rate of de novo synthesis of xanthione oxidase [*Catignani* et al., 1974] in experimental animals. However, the precise role of α-tocopherol in the regulation of protein synthesis is not yet known.

It has been suggested that the antioxidant function of α-tocopherol in vivo may be fortified by the existence of redox couples, that there may be some repair systems working in conjunction with α-tocopherol to exert certain antioxidative effects, and that vitamin E may act as a redox controller of selenium in membrane protein [*Tappel*, 1962; *Diplock*, 1974]. *Nakamura and Hishinuma* [1978] observed that after exhibiting its antioxidant function, almost all α-tocopherol was recovered in unchanged form and suggested that the function of α-tocopherol may be maintained or facilitated by an unknown repair system which reverts the tocopherol radical to α-tocopherol. In an electrochemical study, *Svanholm* et al. [1974] demonstrated that interconversion could take place among α-tocopherol, its cation and its dication, but formation of α-tocopheryl quinone was an irreversible step (fig. 2). *Packer* et al. [1979] have demonstrated the regeneration of α-tocopherol oxidized by carbon trichloroperoxy radical at the expense of vitamin C in an organic solvent. This 'repair system' hypothesis appears to provide an alternative explanation in terms of vitamin E function in vivo between the membrane structure and antioxidant theories. Using an electron spin resonance spectroscopic technique, *Niki* et al. [1982] have also demonstrated that the α-chromanoxy radical formed by the interaction of α-tocopherol with alkoxy radical or 2,2-diphenyl-1-picrylhydrazyl may react with either glutathione or vitamin C (ascorbic acid) to regenerate α-tocopherol. It appears that a hydrogen atom is quickly transferred from glutathione and vitamin C to the α-chromanoxy radical to regenerate α-tocopherol. This type of regeneration may explain at least partly the effective antioxidant effect of α-tocopherol in vivo. However, the precise nature of the 'repair system' of vitamin E in vivo remains to be established.

Plasma

Normal blood volume in men constitutes 6–8%, and in women 5.5–7%, of the total body weight, depending on the bodily constitution. As is shown in table IV, plasma comprises over 50% of whole blood volume. The mean α-tocopherol concentration among normal healthy human subjects averaged approximately 1 mg/dl, and γ-tocopherol ranged from 0.1 to 0.2 mg/dl (table IV). Only a minute amount ($<0.3 \mu g/ml$) of each of α-tocotrienol, β-tocopherol, γ-tocotrienol and *d*-tocopherol, but no δ-tocopherol, was detected in plasma [*Chow*, 1975b]. The levels of α- and γ-tocopherols have been shown to be correlated significantly with the plasma concentration of total cholesterol and triglyceride [*Vatassery* et al., 1983].

Mino et al. [1980, 1982] have shown that plasma in umbilical cord blood and blood of breast-fed infants contain relatively high levels of α- and γ-tocopherols in a ratio resembling that of adults, while that of bottle-fed infants contain relatively high levels of α-tocopherol and some γ-tocopherol. The plasma tocopherol of the bottle-fed infants is rather similar to that of the intake pattern of the average population [*Bieri and Evarts*, 1973; *Chow*, 1975b]. α-Tocopherol levels were also found to be lower in the red cells of bottle-fed infants than in breast-fed infants [*Mino* et al., 1980, 1982]. Whether or not the differences can be attributed to the presence of higher non-α-tocopherols in the plasma of bottle-fed infants competing with α-tocopherol for the available lipoprotein carriers remains to be established.

The distribution of tocopherol has been shown to be related to the lipid content of plasma [*Davies* et al., 1969; *Peake* et al., 1972; *Rubinstein* et al., 1969]. It has been suggested that the concentration of plasma tocopherol be

Table IV. Tocopherol level concentrations in adult human plasma, erythrocytes, leukocytes and platelets

Parameter	% of whole blood	Tocopherol level	*Chow* [1975b][1]	*Bieri* et al. [1979][2]	*Hatam and Kayden* [1979][1]	*Tangney* et al. [1981][1]	*Vatassery* et al. [1983][2]
Plasma, mg/dl	45–65	α-	0.96±0.27	0.95±0.09	0.88±0.34	1.05±0.22	0.99±0.05
		γ-	0.16±0.04	–[3]	0.09±0.02	0.14±0.06	0.21±0.02
Erythrocytes, mg/dl	35–55	α-	0.14±0.03	0.19±0.02	0.14±0.02	–	–
		γ-	0.02±0.01	–	0.02±0.01	–	–
Platelets, μg/10^{10} cells	0.1–0.3	α-	–	–	5.10±0.63	4.04±0.82	3.54±0.13
		γ-	–	–	0.65±0.19	0.96±0.06	0.99±0.08
Leukocytes, μg/10^9 cells	0.03–0.1						
Neutrophils (54–62%)		α-	–	–	4.47±0.62	–	–
		γ-	–	–	0.83±0.22	–	–
Lymphocytes (20–30%)		α-	–	–	3.89±0.05	–	–
		γ-	–	–	0.74±0.23	–	–
Monocytes (5–10%)			–	–	–	–	–
Eosinophils (2–4%)			–	–	–	–	–
Basophils (0.5–1.5%)			–	–	–	–	–

[1] Mean ± standard deviation. [2] Mean ± standard error. [3] Not done or not detectable.

expressed in terms of lipid content, and a ratio of 0.8 mg total tocopherols per gram total plasma lipid would indicate an adequate nutritional status [*Horwitt* et al., 1972]. Based on this ratio, some individuals with plasma total tocopherols below 0.5 mg/100 ml, the commonly accepted satisfactory level, may not be considered as inadequate.

It has been established that tocopherol is associated with lipoproteins in blood plasma (serum) and is transiently present in chylomicrons following a meal. Tocopherol is virtually absent from the plasma of patients with abetalipoproteinemia, a rare impaired absorption disease due to a lack of chylomicron formation and absence of β-lipoprotein [*Kayden and Silber*, 1965]. Other conditions such as premature infancy and cystic fibrosis (the most common cause of pancreatic insufficiency during the first three decades of life), in which less than normal serum tocopherol concentrations have been reported, may also be due to a decreased availability of β-lipoprotein as a carrier. *Kater* et al. [1970] have shown a marked decrease in serum tocopherol concentrations in patients with severe cirrhosis and fulminant hepatitis. They have also observed a positive close correlation between the tocopherol concentration and the β-lipoprotein level in the sera from all groups of patients and controls. Thus, low serum tocopherol levels in patients with cirrhosis and fulminant hepatic failure may partially result from impairment of β-lipoprotein synthesis by the liver. This finding also suggests that serum tocopherol levels may depend on the availability of a carrier for transport rather than tocopherol absorption or availability in the diet. However, the importance of impaired tocopherol transport in the initiation of erythrocyte hemolysis and of tissue tocopherol deficiency associated with malnutrition, premature infancy, malabsorption, liver disease, abetalipoproteinemia and other tocopherol-associated syndromes remains to be established.

The relative amounts of α-tocopherol and its isomers contained in different fractions of lipoproteins have been a subject of controversy. For example, *McCormick* et al. [1960] and *Pelkonen* [1960] reported that most of the serum tocopherol was in the low-density lipoprotein, while *Lewis* et al. [1954] and *Takahashi* et al. [1977] reported that high-density lipoprotein was the main carrier of the vitamin. *Behrens* et al. [1982] reinvestigated the distribution of α-tocopherol in plasma lipoproteins of humans of both sexes and concluded that tocopherol and protein levels were highly correlated in high-density lipoprotein, rather than the lower density lipoprotein. They also observed a sex-dependent difference of α-tocopherol distribution in plasma lipoprotein fractions, and that the difference was due to the different levels of proteins in those lipoprotein fractions. More studies are needed to have a better understanding of the role of various fractions of lipoproteins in the absorption and turnover of tocopherols.

Erythrocytes (Red Blood Cells)

Red blood cells comprise 41–52% of normal blood volume in men, and about 35–48% in women. α-Tocopherol concentration in the red cells has been reported to be approximately 15–25% (0.5–2.0 µg/ml) that of the blood plasma (table V). γ-Tocopherol concentration in the red cells ranges from 0.1 to 0.4 µg/ml. Except for a tiny amount of δ-tocopherol, no other isomers have been detected in the red cells [*Chow*, 1975b; *Tangney* et al., 1981]. The red cell tocopherol has been shown to be in a dynamic equilibrium with its counterpart in the plasma, and a relatively narrow ratio of red cells to plasma tocopherol is maintained [*Silber* et al., 1969]. Specific binding sites for α-tocopherol in human erythrocytes have also been shown to exist [*Kitabchi and Wimalasena*, 1982].

While subserving biological gas exchange, erythrocytes may come under various oxidative stresses, such as due to inhaled oxidant gases and ingested oxidant drugs. However, the metabolic machinery of the red cells is able to reverse most of the adverse effects under normal conditions. Only when the reducing power of the red cells to reverse oxidative injury is decreased, or when the oxidative stress is exceedingly increased, may irreversible damage to the red cells occur. Like other parenchyma organs, red blood cells contain a variety of enzymatic and non-enzymatic systems which may be protective against oxidative damage. Among the important antioxidant systems of the red cells (fig. 3), vitamin E appears to be the most important and dominant one [*Chow*, 1979].

While it is well-known that vitamin E-deficient erythrocytes are more susceptible to lysis by hemolytic agents, no apparent structural differences between the membrane proteins of vitamin E-deficient and normal cells have been revealed. Experimental evidence available suggests that reduced levels of dietary vitamin E increase the ability of the erythrocyte membrane to serve as a substrate for the oxidative formation of disulfide bonds between proteins not normally covalently linked [*Sayare* et al., 1982]. Also, membranes from vitamin E-deficient cells have a greater propensity to cross-link hemoglobin to band 3 than do membranes from normal cells [*Sayare* et al., 1982]. Because of its membrane localization property, one

Table V. Effect of vitamin E administration on human polymorphonuclear cell metabolism and function[1]

Parameter	Control	Vitamin E	p
Oxygen utilization[2]	61 ± 4	66 ± 2	0.12
Superoxide production[2]	5 ± 2	7 ± 2	0.29
Hydrogen peroxide release[2]	3 ± 1	0.1 ± 0.3	0.015
HMP shunt activity[3]			
Resting	67 ± 19	38 ± 10	0.20
Phagocytosing	299 ± 42	113 ± 8	0.025
Release of arachidonic acid[4]			
Resting	271 ± 38	424 ± 25	0.02
0.7 mg zymosan	950 ± 180	$1,462 \pm 320$	0.02
1.4 mg zymosan	$1,943 \pm 230$	$2,370 \pm 811$	0.50

[1] From *Baehner* et al. [1982]. Vitamin E given at 1,600 U/day for at least 2 weeks.
[2] Nanomoles/min/10^7 PMNs. Values are the means and standard deviations of replicate samples.
[3] Micromoles $^{14}CO_2$/30 min/10^7 PMNs.
[4] dpm/5 min/10^7 PMNs.

conceivable biochemical role of vitamin E is to terminate free radical-generated lipid peroxidation chain reactions, particularly in membranes that are rich in polyunsaturated lipids. This is consistent with the findings that the cell membrane is the primary site for the observed damage associated with a vitamin E deficiency [*Molenaar* et al., 1970].

Hemoglobin is known to catalyze the process of lipid peroxidation through the decomposition of lipid hydroperoxides to free radicals and other reactive oxygen species. Auto-oxidation of oxyhemoglobin to methemoglobin results in the generation of superoxide radicals [*Weaver* et al., 1973; *Koppenol and Butler,* 1977]. The reaction of a superoxide radical with hydrogen peroxide in the red cells may produce such highly reactive intermediates as a hydroxy radical and a singlet oxygen. These reactive oxygen species in turn may react with the lipid and protein components, especially polyunsaturated phospholipids and sulfhydryl compounds of the membrane and, as a result, may alter the membrane's permeability and functions.

Destruction of specific membrane phospholipids during peroxidative hemolysis of vitamin E-deficient erythrocytes has been demonstrated

[*Heikkila* et al., 1971]. Any disruption of the integrity of the native membrane structure will consequently affect important biological processes in the cells. In addition to lipid peroxidation, proteins interspaced in the bilayer may possibly be altered by free radicals originating in the lipid phase of the membrane. Furthermore, there are structural proteins arranged in a filamentous network on the cytoplasmic face of the membrane. This network is responsible for supporting and maintaining the intact bilayer. Because of the proximity of these structural proteins to the membrane, these proteins may also be modified by free radicals present in the phospholipid bilayer or form cross-linking compounds with such lipid peroxidation products as malonaldehyde. Cumulative damage of the membrane integrity may eventually lead to hemolysis of the red cells. *Shapiro and Mott* [1982] have provided experimental evidence that the sites of membrane damage in vitamin E-deficient erythrocytes are on the cytoplasmic face of the membrane, closer to the source of the reactive radical species. Following a quantitative comparison of the vitamin E to total antioxidant concentrations, *Burton* et al. [1983] concluded that vitamin E (α-tocopherol) is the major, and probably the only, lipid-soluble, free radical chain-breaking antioxidant in adult human blood plasma and red blood cells. They also observed that a large supplement of vitamin E did not significantly increase the vitamin E-to-lipid ratio in plasma or red blood cells.

While evidence available indicates that vitamin E is a specific erythropoietic factor for nonhuman primates and swine, vitamin E has normally not been shown to be required as an erythropoietic factor for humans or several other species of animals [*Baker* et al., 1968; *Fitch,* 1972]. *Drake and Fitch* [1980] suggested that the lack of a requirement for vitamin E in erythropoiesis in humans may be due to a metabolic adaptation that circumvents the need for the role that the vitamin would otherwise serve. Patients with protein-calorie malnutrition who receive vitamin E supplementation have been shown to respond with reticulocytosis and an increase in hemoglobin concentration before their metabolic derangement is reversed by correcting their other nutritional deficiencies. Based on this finding, *Drake and Fitch* [1980] predict that other acquired or congenital abnormalities of metabolism could impair the adaptation that circumvents the role of vitamin E in erythropoiesis. More studies are needed to substantiate this view.

Human adults maintained on a low vitamin E diet for a long period of time have not shown clinical manifestations, although erythrocytes of both adults and infants with low serum tocopherol levels are susceptible to oxidizing agents in vitro, and erythrocyte survival is shorter in these adults than in normal persons [*Horwitt* et al., 1963]. The life span of erythrocytes in full-term infants is approximately two thirds that of the erythrocytes of normal adults. The erythrocytes of premature infants have an even shorter life span than do those of full-term infants [*O'Brien and Pearson,* 1971]. Deficiency of vitamin E in premature infants at birth may at least partially play a role in this accelerated senescence [*Gross* et al., 1977]. This has been shown by the partial inhibition of the accelerated red cell destruction when vitamin E was administered to premature infants. Infants, especially those of low birth weight, have been repeatedly shown to have low serum tocopherol levels [*Moyer,* 1950; *Wright* et al., 1951; *Nitowsky* et al., 1956]. In premature infants, such syndromes as hemolytic anemia reticulocytosis, thrombocytosis and edema have been shown to be responsive to oral administration of vitamin E [*Oski and Barnes,* 1967; *Ritchie* et al., 1968]. Low birth weight infants fed formulas containing low ratios of vitamin E to polyunsaturated fatty acids and added iron may be receiving an inadequate amount of vitamin E, and the formulas may be a contributing factor in the cause of anemia, edema and other vitamin E deficiency syndromes [*Hassan* et al., 1966; *Ritchie* et al., 1968].

In addition to abetalipoproteinemia, chronic steatorrhea, such as cystic fibrosis, and premature infancy, low serum vitamin E levels in humans have also been reported to be associated with congenital hemolytic anemias, such as sickle-cell anemia (SCA), β-thalassemia and glucose-6-phosphate dehydrogenase (G6PD) deficiency. The cause for vitamin E deficiency in chronic steatorrhea is related to inadequate absorption due to physiological or pathological changes in the gastrointestinal tract. In the cases of congenital hemolytic anemia, vitamin E deficiency most likely results from an increased consumption of the vitamin due to increased oxidative stress in pathological membranes and in other tissues.

SCA is a genetic disorder caused by a point mutation in DNA that codes for valine rather than glutamic acid in the sixth position of the β-globin chain of the hemoglobin tetramer. The erythrocytes of SCA patients are more susceptible to lipid peroxidation than are normal erythrocytes. Decreased blood levels of vitamin E and abnormal membrane phospholipid organization induced by sickling may be partially responsible for the increased susceptibility of the sickled cells to peroxidative damage [*Chiu* et al., 1979,1982; *Lubin and Chiu,* 1982]. The peroxidative damage of sickled erythrocytes may accelerate or contribute to loss of cell deformability and to chronic hemolysis. These processes may then lead to the production of abnormal cellular properties, such as potassium leak and reduced filterability, and contribute to the formation of irreversibly sickled cells. Supplementation of vitamin E has been shown to have a beneficial effect on SCA patients. A scheme to describe the role of lipid peroxidation and vitamin E in the pathophysiology of SCA has been proposed by *Chiu* et al. [1982].

G6PD deficiency, a sex-linked trait, has been reported worldwide and exhibits extensive genotypic and phenotypic polymorphism [*Beutler,* 1978]. G6PD-deficient red cells generate abnormally small quantities of reduced nucleotides in response to oxidant stress which results in an increased sensitivity to the adverse effects of oxidant exposure [*Luzzato and Testa,* 1978].

Erythrocyte GSH synthetase deficiency is a less common cause of chronic hemolytic anemia and may occur with or without oxoprolinuria, depending upon the specific nature of the enzyme deficiency [*Spielberg* et al., 1978]. Individuals with this disorder exhibit a significant reduction in red cell survival and moderately severe anemia [*Mohler* et al., 1970]. Decreased levels of GSH due to inadequate synthesis render the red cells susceptible to oxidant stress. In addition to being the key substrate of GSH peroxidase, GSH has other important functions of its own [*Chow,* 1979]. As in the case of G6PD-deficient patients, the red cells of GSH synthetase-deficient patients are more susceptible to oxidative stress. Treatment of the patients with vitamin E has been shown to improve both red cell life span and polymorphonuclear cell function [*Corash* et al., 1982]. Vitamin E treatment, however, has not been shown to improve the red cell survival or membrane protein pattern of G6PD-deficient patients [*Corash* et al., 1982; *Johnson* et al., 1981, 1983].

Several intracellular changes such as excess α-hemoglobin subunits and intracellular iron in thalassemic erythrocytes could account for the oxidative damage in their membrane components [*Rachmilewitz* et al., 1982]. Due to the decreased amounts of intracellular hemoglobin that can serve as a substrate for the active oxygen radicals, the free radical formed may have an easier access to the cell membrane and thus oxidize the cellular components. However, while indirect evidence for the deleterious role of the free radicals has been obtained from the studies of the lipid membranes of thalassemic erythrocytes, no direct evidence of excess free oxygen radical formation in the red cells has been reported [*Rachmilewitz* et al., 1976].

Plasma levels of vitamin E have also been found to be decreased (0.5 mg/dl) in patients with thalassemia or Gaucher's disease. However, there is no evidence for intestinal malabsorption of the lipid-soluble vitamins in these patients. Rapid consumption of vitamin E appears to occur concurrently while neutralizing the oxidative damage in the pathological erythrocyte membranes and in other tissues in patients with thalassemia. Lysosomal accumulation of glucocerebroside may stimulate phagocytes into a maintained 'respiratory burst' with excessive production of oxygen free radicals, resulting in increased utilization and eventual deficiency of vitamin E in patients with Gaucher's disease [*Rachmilewitz* et al., 1982].

Leukocytes (White Blood Cells)

Leukocytes may be classified as being either granular or nongranular. Granulocytes have an abundance of granules in the cytoplasm and may be divided into three groups: (a) neutrophils, cells with fine neutrophilic granules in the cytoplasm and an irregular lobed nucleus; (b) eosinophils, cells with coarse eosinophilic granules in the cytoplasm and a lobed nucleus, and (c) basophils, cells with coarse basophilic granules in the cytoplasm and a bent nucleus that is partially constricted into two lobes. Nongranulocytes lack specific granules in their cytoplasm which includes the lymphocytes and monocytes. Leukocytes comprise less than 0.1% of blood volume in humans. Of the total number of leukocytes, more than half are neutrophils and approximately one quarter are lymphocytes. The concentrations of α- and γ-tocopherols in neutrophils and lymphocytes are shown in table IV. The tocopherol content in the white cells is relatively higher than that of the red cells and platelets. For example, the per cell value of α-tocopherol content of neutrophils and lymphocytes is 10-fold greater than that of platelets, and 35-fold greater than the red cell (table IV).

Eosinophils and basophils are the minor constituents of the leukocytes, and relatively little information is available concerning these leukocytes. The number of eosinophils and neutrophils was found to be significantly elevated in the vitamin E-deficient rats relative to the supplemented group. However, the number of lymphocytes, but not monocytes or total leukocyte count, was significantly decreased in the deficient group [*Chow* et al., 1981]. The significance of this difference is not yet clear.

Polymorphonuclear leukocytes (neutrophils) and monocytes comprise the blood phagocytes. These cells, unlike the other cellular elements of the blood, are capable of leaving the circulation and migrating to extravascular sites of infection and inflammation where they recognize and ingest microbes and other particulate material. Similar to platelets and lymphocytes but unlike red blood cells, neutrophils and monocytes are capable of achieving a heightened state of functional and metabolic activation in response to immunochemical and particulate signals. The metabolic activation process includes activation of one or more cyanide-insensitive membrane oxidases and related enzyme systems, which oxidize substrate NADH and NADPH. This leads to increased oxygen utilization and the subsequent reduction of oxygen to superoxide anions, hydrogen peroxide, singlet oxygen, hydroxy radicals and other free radicals [*Baehner* et al., 1982; *Fantone and Ward,* 1982]. It has also been shown that these reactive oxygen species, especially hydrogen peroxide, are involved in intracellular microbicidal reactions linked to the myeloperoxidase halide system present in neutrophils and monocytes [*Klebanoff and Clark,* 1978].

While the requirement of oxygen in the bacterial killing is critical, the reactive oxidative oxygen species may also play a role in modulating the cytotoxicity of the host cells [*Fantone and Ward,* 1982]. Furthermore, the lethal reactions are potentiated by catalytic peroxidative reactions involving granula myeloperoxidase working in concert with hydrogen peroxide and a halide, either in the form of chloride or iodide [*Klebanoff,* 1968]. In addition to halogenation, the myeloperoxidase system is capable of decarboxylation of amino acids, converting them to carbon dioxide and ammonia [*Straus* et al., 1970]. These reactions occur not only on bacterial membranes, but also on the membranes of the phagocytic cell itself.

Similar to erythrocytes and other tissues, neutrophils and monocytes possess antioxidant defense systems for their protection against the deleterious effects of reactive oxygen species generated (fig. 3). Hydroperoxides formed can be reduced through the action of the GSH peroxidase system. The restoration of GSH involves the reaction of oxidized glutathione (GSSG) through the action of GSH reductase in which NADP is regenerated. The latter in turn leads to the activation of the HMP shunt through G6PD to provide a continued supply of NADPH. Catalase, which is also present in the phagocytes, is effective only at high concentrations of hydrogen peroxide, whereas GSH peroxidase is effective at physiologic concentrations of hydrogen peroxide. By employing discriminant reactive oxygen species scavengers, *Baehner* et al. [1977] found that hydrogen peroxide, not the superoxide anion, hydroxy radical, or singlet oxygen, was responsible for modulating the phagocytic rate. Similarly, hydrogen peroxide was found to attenuate chemotaxis of the neutrophils.

Because the cellular functions of chemotaxis and ingestion require membrane recognition, it is likely that the effect of the hydrogen peroxide may be directed primarily to the surface membrane. For example, it has been shown that the complement C_3 receptor compared to the immunoglobulin F_c receptor of the neutrophil is particularly vulnerable to both superoxide anion and hydrogen peroxide, resulting in impaired binding of C_3-coated bacteria to the membrane [*Boxer* et al., 1979]. In a study to determine the role of vitamin E in neutrophil function, *Baehner* et al. [1977] observed that vitamin E-repleted neutrophils consistently ingested at an accelerated rate compared to pretreatment values. The bactericidal capacity of the vitamin E-repleted neutrophils was consistently decreased to 70% of control values. The release of superoxide anion from phagocytizing neutrophils was unaltered, while less hydrogen peroxide was elaborated in comparison to control samples. The findings suggest that vitamin E may selectively scavenge hydrogen peroxide and not superoxide anion, and that direct movement and phagocytosis by neutrophils can be attenuated by peroxidative damage to the cell membrane.

Using rats as an experimental model to study the effects of vitamin E deficiency on neutrophil and monocyte function, *Harris* et al. [1980] have shown that daily administration of large amounts of vitamin E resulted in alterations of the ability of human neutrophils to ingest and kill bacteria, to utilize oxygen, to release the oxygen by-products superoxide anion and hydrogen peroxide, and to release arachidonic acid presumably through activation of membrane phospholipase A. Changes in membrane fluidity of normal and vitamin E-supplemented cells at rest and during phagocytosis also occurred. The NADH and NADPH oxidase activities in normal and vitamin E-supplemented cells could be localized to plasma membranes and phagocytic vesicles. Thus, vitamin E appears to provide a probe for differentiating the metabolic and biologic role of each oxidase.

Insufficient quantities of vitamin E have also been shown to be directly associated with impaired neutrophil response [*Harris* et al., 1978]. Neutrophils and peritoneal macrophages obtained from rats maintained on a vitamin E-deficient diet for 2 months have been shown to have impaired chemotaxis of 68 and 44% of control values, respectively. The ingestion of complement C_3-coated particles was decreased to 44% in both cell types. The vitamin-depleted cells also secreted lysosomal enzyme, β-glucuronidase, into the extracellular fraction normally, and sustained lipid peroxidation damage. These findings support the view that vitamin E is required for the maintenance of normal membrane functions of direct movement and ingestion by phagocytic cells.

Platelets

Platelets are disk-shaped structures, 2–4 μm in diameter, found in the blood of all mammals, and are known for their role in blood coagulation. Platelets comprise approximately 0.1–0.3% of whole blood volume. The mean concentration of platelet tocopherol has been reported to be 4–5 μg/10^{10} platelets for α-tocopherol, and up to 1 μg/10^{10} platelet for the γ-form (table IV). Unlike those in the plasma, the platelet α- and γ-tocopherol concentrations do not show significant correlation with plasma lipid, cholesterol or triglyceride concentrations. Since the platelet tocopherol concentrations do not positively reflect the plasma lipid concentrations, it has been suggested that the concentration of platelet tocopherol may serve as a more accurate indicator of vitamin E nutritional status than that of the plasma tocopherol [*Vatassery* et al., 1983].

Platelets have a relatively high content of polyunsaturated fatty acid, particulary in the granula and membrane phospholipids [*Nordoy* et al., 1969]. In addition to protecting polyunsaturated fatty acids in phospholipids from oxidation, vitamin E reduces the permeability of membranes that contain highly unsaturated phospholipids, and protects the active membrane sites from phospholipase [*Lucy,* 1972]. An adverse effect on platelets may therefore result from a vitamin E deficiency and a high intake of dietary polyunsaturated fatty acids. *Nordoy and Strom* [1975] have shown that the human platelets contain about 3 times as much total tocopherol per cell as erythrocytes, and the difference is not related to the polyunsaturated fatty acid content. They also observed that unlike that of plasma and erythrocytes, the tocopherol level in platelets was not altered significantly by tocopherol treatment or by in vitro incubation. Platelet aggregation is an oxygen-requiring process. It is associated with a burst in oxygen consumption needed principally for the metabolism of arachidonic acid liberated from membrane phospholipid and is inducible by certain agents [*Hussain and Newcomb,* 1964; *Pickett and Cohen,* 1976]. Lipid hydroperoxide, a by-product of this process, has been recognized as an essential feature of aggregation [*Okuma* et al., 1971].

There has been an interest in determining the role of vitamin E in platelet function, especially as a potential antiaggregating agent of platelets. Abnormalities in platelet function or platelet-vascular interactions have been demonstrated to occur in several conditions associated with low plasma levels of vitamin E. Patients with an impairment in fat absorption (i.e. cystic fibrosis, a betalipoproteinemia) who have not been supplemented with vitamin E show enhanced platelet aggregation and malonaldehyde formation, and in some cases a decrease in platelet life span [*Stuart,* 1982]. Following vitamin E supplementation, platelet aggregation and survival measurements have been shown to return to normal, suggesting a cause-and-effect relationship. While vitamin E has been shown to inhibit platelet aggregation, platelet release, and malonaldehyde formation in vitro, attempts to demonstrate corresponding effects in vivo have not been conclusive [*Steiner and Anastasi,* 1976; *Stuart and Oski,* 1979]. Also, the platelet lipid metabolism and function have not been found to be altered significantly by a high intake of vitamin E for a prolonged period of time [*Nordoy and Strom,* 1975].

Lake et al. [1977] demonstrated the presence of platelet hyperaggregability to adenosine diphosphate, epinephrine, and collagen in infants with vitamin E deficiency. They also showed that a complete reversal of the platelet abnormalities occurred following the attainment of normal vitamin E levels. Using rats as an animal model, *Machlin* et al. [1975] have shown that collagen-induced platelet aggregation was increased and serum level of PGE_2 and $PGF_{2\alpha}$ were elevated in the vitamin E-deficient animals. The findings are consistent with observations of increased production of platelet prostaglandin endoperoxide in the vitamin E-deficient platelets [*Stuart,* 1982; *Lake* et al., 1977] and suggest an enhancement of the platelet cyclooxygenase pathway in vitamin E deficiency. However, since both vitamin E and its quinone have been shown to be equally effective in inhibiting cyclooxygenase over the lipoxygenase pathway and cAMP phosphodiesterase [*Steiner and Mower,* 1982], an alternative explanation to the antioxidant theory of the vitamin E effect needs to be sought. It has been suggested that the effect of vitamin E on membrane fluidity may form the basis of vitamin E action in platelets [*Steiner,* 1981; *Steiner and Mower,* 1982].

Recently, *Steiner* [1983] reported that in women, vitamin E administration (up to 1,200 IU) produced a small but significant reduction of collagen-induced platelet aggregation. A similar trend was also seen in men. Adhesiveness of platelets to collagen was not affected by aspirin ingestion but showed a highly significant reduction in vitamin E and vitamin E plus aspirin-treated individuals. The efficiency of vitamin E as a potential inhibitor of platelet function suggests that vitamin E administration could have a beneficial effect in patients suffering from arterial thromboembolic diseases. Further studies are needed to have a better understanding of the role of vitamin E in platelet function and its potential applications to platelets associated with clinical manifestations.

Conclusions

More than 50 years have passed since vitamin E was originally described as a fat-soluble dietary substance necessary for reproduction in rats. Since the Food and Nutrition Board of the US National Research Council included vitamin E as a recommended nutrient for man in 1965, it has remained unique among the vitamins in lacking an unequivocal deficiency state in man. While it is difficult to produce vitamin E deficiency in human adults [*Horwitt*, 1962], vitamin E deficiency can be induced rather easily in many laboratory animals. A wide variety of lesions have been produced in animals ranging from encephalomalacia to hepatic or renal degeneration and nutritional muscular dystrophy [*Scott*, 1969]. A number of biochemical abnormalities, such as the increased susceptibility of red blood cells to hemolytic agents, and leakage of tissue enzymes into blood circulation, can be readily detected in vitamin E-deficient animals. However, due to the lack of definite correlation between clinical deficiency syndromes attributed to the vitamin E deficiency in man, vitamin E has been termed as 'a vitamin in search of a disease', and effort to understand vitamin E has been described as 'a bewildering prospect because of its protean nature' by many clinicians.

On the other hand, there is a continued interest in the potential role of vitamin E as an important agent in human cellular metabolism. In addition to its ability in alleviating deficiency symptoms of experimental animals, dietary vitamin E has been shown to modify the toxicity or cellular responses/susceptibility of such a variety of compounds as ozone [*Chow*, 1982], oxygen [*Mino*, 1973], paraquat [*Block*, 1979], lead [*Levander* et al., 1977], nitrofurantoin [*Boyd* et al., 1979], methyl mercury [*Welsh*, 1979], adriamycin [*Doroshow*, 1979], dimethylnitrosamine [*Dashman and Kamm*, 1979] and nitrite [*Chow* et al., 1981]. A role in modifying the acute respiratory-distress and development of bronchopulmonary dysplasia in the human newborn has also been reported [*Ehrenkranz* et al., 1982]. These findings suggest a need or use of this vitamin. Furthermore, vitamin E has been recommended for treatment of such illnesses as impotence, muscular dystrophy and aging without the backing of sound experimental or clinical studies. Due in part to considerable tissue storage, it is difficult to produce a vitamin E deficiency under experimental conditions in adult man, yet deficiencies or subclinical deficiencies do occur in association with intestinal malabsorption syndromes of various etiologies [*Bieri and Farrell*, 1976]. The findings that vitamin E deficiencies or the needs to supplement vitamin

E in some infants and in adults with fat absorption problems clearly indicate a role of vitamin E in human nutrition. A better understanding of its precise role in human nutrition and health is urgently needed.

References

Arimori, S.; Sumitomo, K.: Ultrastructure of platelets of the vitamin E-deficient rats. J. Nutr. Sci. Vitaminol. *23:* 372–384 (1977).
Baehner, R.L.; Boxer, L.A.; Allen, J.M.; Davis, J.: Autooxidation as a basis for altered function by polymorphonuclear leukocytes. Blood *50:* 327–335 (1977).
Baehner, R.L.; Boxer, L.A.; Ingraham, L.M.; Butterick, C.; Haak, R.A.: The influence of vitamin E on human polymorphonuclear cell metabolism and function. Ann. N.Y. Acad. Sci. *393:* 237–250 (1982).
Baker, S.J.; Pereira, S.M.; Begum, A.: Failure of vitamin E therapy in the treatment of anemia of protein-calorie malnutrition. Blood *32:* 717–725 (1968).
Barnes, P.J.; Taylor, P.W.: Developments in tocopherol analysis. Chem. New Engl. *45:* 176–180 (1981).
Bauernfeind, J.C.: The tocopherol content of food and influencing factors. CRC crit. Rev. Food Sci. Nutr. *8:* 337–382 (1977).
Bauernfeind, J.C.; Cort, W.M.: Tocopherols; in Encyclopedia of food technology, pp. 891–899 (AVI Publishing, Westport 1974).
Behrens, W.A.; Thompson, J.N.; Madere, R.: Distribution of α-tocopherol in human plasma lipoproteins. Am. J. clin. Nutr. *35:* 691–696 (1982).
Beutler, E.: Hemolytic anemia in disorders of red cell metabolism (Plenum Publishing, New York 1978).
Bieri, J.G.: Kinetics of tissue α-tocopherol depletion and repletion. Ann. N.Y. Acad. Sci. *203:* 181–191 (1972).
Bieri, J.G.; Evarts, R.P.: Tocopherols and fatty acids in American diets. J. Am. diet. Ass. *62:* 147–151 (1973).
Bieri, J.G.; Evarts, R.P.: Vitamin E activity of gamma-tocopherol in the rat, chick and hamster. J. Nutr. *104:* 850–857 (1974).
Bieri, J.G.; Farrell, P.M.: Vitamin E. Vitams Horm. *34:* 31–75 (1976).
Bieri, J.G.; Poukka, R.K.H.: Red cell content of vitamin E and fatty acids in normal subjects and patients with abnormal lipid metabolism. Int. Z. VitamForsch. *40:* 344–350 (1970).
Bieri, J.G.; Tolliver, T.J.; Catignani, G.L.: Simultaneous determination of alpha-tocopherol and retinol in plasma or red blood cells by high pressure liquid chromatography. Am. J. clin. Nutr. *32:* 2143–2149 (1979).
Block, T.R.: Potentiation of acute paraquat toxicity by vitamin E deficiency. Lung *156:* 195–203 (1979).
Blomstrand, R.; Forsgren, L.: Labelled tocopherols in man. Intestinal absorption and thoracic-duck lymph transport of *dl*-alpha-tocopherol-3,4-^{14}C2 acetate, *dl*-alpha-tocopheramine-3,4-^{14}C2, *dl*-alpha-tocopherol-(5-methyl-^{3}H) and *N*-(methyl-^{3}H)-*dl*-γ-tocopheramine. Int. Z. VitamForsch. *38:* 328–344 (1968).
Boxer, L.A.; Oliver, J.M.; Spielberg, S.P.; Allen, J.M.; Schulman, J.P.: Protection of gran-

ulocytes by vitamin E in glutathione synthetase deficiency. New Engl. J. Med. *301:* 901–905 (1979).

Boyd, M.R.; Catignani, G.L.; Sasame, H.A.; Mitchell, J.R.; Stiko, A.W.: Acute pulmonary injury in rats by nitrofurantoin and modification by vitamin E, dietary fat and oxygen. Am. Rev. resp. Dis. *120:* 93–99 (1979).

Burton, G.W.; Joyce, A.; Ingold, K.U.: Is vitamin E the only lipid-soluble, chain breaking antioxidant in human blood plasma and erythrocyte membranes? Archs Biochem. Biophys. *221:* 281–290 (1983).

Catignani, G.L.; Chytil, F.; Darby, W.J.: Vitamin E deficiency: immunochemical evidence for increased accumulation of liver xanthione oxidase. Proc. natn. Acad. Sci. USA *71:* 1966–1968 (1974).

Chiu, D.; Lubin, B.; Shohet, S.B.: Erythrocyte membrane lipid reorganization during the sickling process. Br. J. Haematol. *41:* 223–234 (1979).

Chiu, D.; Vichinsky, E.; Yee, M.; Kleman, K.; Lubin, B.: Peroxidation, vitamin E, and sickle-cell anemia. Ann. N.Y. Acad. Sci. *393:* 323–335 (1982).

Chow, C.K.: Increased activity of pyruvate kinase in plasma of vitamin E-deficient rats. J. Nutr. *105:* 1221–1224 (1975a).

Chow, C.K.: Distribution of tocopherols in human plasma and red blood cells. Am. J. clin. Nutr. *28:* 756–760 (1975b).

Chow, C.K.: Nutritional influence on cellular antioxidant defense systems. Am. J. clin. Nutr. *32:* 1066–1081 (1979).

Chow, C.K.: Influence of dietary vitamin E on susceptibility to ozone exposure. Adv. Mod. Envir. Toxicol. *5:* 75–93 (1982).

Chow, C.K.; Csallany, A.S.; Draper, H.H.: Turnover rates of tocochromanols in rabbit plasma. Nutr. Rep. int. *4:* 45–48 (1971).

Chow, C.K.; Draper, H.H.; Csallany, A.S.; Chiu, M.: The metabolism of C-14-alpha-tocopheryl quinone and C-14-alpha-tocopheryl hydroquinone. Lipids *2:* 390–396 (1967).

Chow, C.K.; Hong, C.B.; Gairola, C.; Reese, M.: Potentiation of nitrite toxicity by vitamin E deficiency. Fed. Proc. *40:* 874A (1981).

Chow, C.K.; Tappel, A.L.: An enzymatic protective mechanism against lipid peroxidation damage to lungs of ozone-exposed rats. Lipids *7:* 518–524 (1972).

Corash, L.M.; Sheetz, M.; Bieri, J.G.; Bartsocas, C.; Moses, S.; Bashan, N.; Schulman, J.D.: Chronic hemolytic anemia due to glucose-6-phosphate dehydrogenase deficiency or glutathione synthetase deficiency: the role of vitamin E in its treatment. Ann. N.Y. Acad. Sci. *39:* 348–360 (1982).

Csallany, A.S.; Draper, H.H.: Dimerization of alpha-tocopherol in vivo. Archs Biochem. Biophys. *100:* 335–337 (1963).

Csallany, A.S.; Draper, H.H.; Shah, S.N.: Conversion of d-α-tocopherol-C^{14} to tocopheryl-p-quinone in vivo. Archs Biochem. Biophys. *98:* 142–154 (1962).

Dashman, T.; Kamm, J.J.: Effects of high doses of vitamin E on dimethylnitrosamine hepatotoxicity and drug metabolism in the rats. Biochem. Pharmacol. *28:* 1485–1490 (1979).

Davies, T.; Kelleher, J.; Losowsky, M.S.: Interrelation of serum lipoprotein and tocopherol levels. Clinica chim. Acta *24:* 431–436 (1969).

Diplock, A.T.: Possible stabilizing effect of vitamin E on microsomal, membrane-bound, selenide-containing proteins and drug-metabolizing enzyme systems. Am. J. clin. Nutr. *27:* 995–1004 (1974).

Doroshow, J.H.; Locker, G.Y.; Myers, C.E.: Experimental animal models of adriamycin cardiotoxicity. Cancer Treat. Rep. *63:* 855–860 (1979).

Drake, J.R.; Fitch, C.D.: Status of vitamin E as an erythropoietic factor. Am. J. clin. Nutr. *33:* 2386–2393 (1980).

Draper, H.H.; Csallany, A.S.; Chiu, M.: Isolation of a trimer of α-tocopherol from mammalian liver. Lipids *2:* 47–54 (1967).

Draper, H.H.; Csallany, A.S.; Shah, S.N.: Isolation and synthesis of a new metabolite of α-tocopherol. Biochim. biophys. Acta *59:* 527–529 (1962).

Draper, H.H.; Johnson, B.C.: *NN*-Diphenyl-*p*-phenylene diamine in the prevention of vitamin E deficiency in the lamb. J. Anim. Sci. *15:* 1154–1160 (1954).

Ehrenkranz, R.A.; Ablow, R.C.; Warshaw, J.B.: Effect of vitamin E on the development of oxygen-induced lung injury in neonates. Ann. N.Y. Acad. Sci. *393:* 452–466 (1982).

Fitch, C.D.: The hematopoietic system in vitamin E-deficient animals. Ann. N.Y. Acad. Sci. *203:* 172–176 (1972).

Fitch, C.D.; Broun, G.O., Jr.; Chou, A.C.; Gallagher, N.I.: Abnormal erythropoiesis in vitamin E-deficient monkeys. Am. J. clin. Nutr. *33:* 1251–1258 (1980).

Fantone, J.C.; Ward, P.A.: Role of oxygen-derived free radicals and metabolites in leukocyte-dependent inflammatory reactions. Am. Soc. Pathol. *107:* 397–418 (1982).

Gabriel, E.; Machlin, L.J.; Filipski, R.; Nelson, J.: Influence of age on the vitamin E requirement for resolution of necrotizing myopathy. J. Nutr. *110:* 1372–1379 (1980).

Gloor, J.; Wursch, J.; Schwieter, U.; Wiss, O.: Resorption, Retention, Verteilung und Stoffwechsel des *dl*-Alpha-Tocopheramins, *d-N*-Methyl-α-Tocopheramins und des α-Tocopherols im Vergleich zum *dl*-α-Tocopherol bei der Ratte. Helv. chim. Acta *49:* 2303–2312 (1966).

Green, J.: Vitamin E and the biological antioxidant theory. Ann. N.Y. Acad. Sci. *203:* 29–44 (1972).

Gross, S.J.; Landaw, S.A.; Oski, F.A.: Vitamin E and neonatal hemolysis. Pediatrics, Springfield *59:* 995–997 (1977).

Harris, P.L.; Embree, N.D.: Quantitative consideration of the effect of polyunsaturated fatty acid content of the diet upon the requirements for vitamin E. Am. J. clin. Nutr. *13:* 385–392 (1963).

Harris, R.E.; Boxer, L.A.; Baehner, R.L.: Abnormal membrane function of phagocytes from vitamin E-deficient rats. Pediat. Res. *12:* 464a (1978).

Harris, R.E.; Boxer, L.A.; Baehner, R.L.: Consequences of vitamin E deficiency on the phagocytic and oxidative function of the rat polymorphonuclear leukocyte. Blood *55:* 338–343 (1980).

Hassan, H.; Hashim, S.A.; Van Itallie, T.B.; Sebrell, W.H.: Syndrome in premature infants associated with low plasma vitamin E levels and high polyunsaturated fatty acid diet. Am. J. clin. Nutr. *19:* 147–157 (1966).

Hatam, L.J.; Kayden, H.J.: A high-performance liquid chromatographic method for the determination of tocopherol in plasma and cellular elements of the blood. J. Lipid Res. *20:* 639–645 (1979).

Hatam, L.; Kayden, H.J.: Tocopherol levels in needle aspiration biopsies of adipose tissue: normal subjects and abetalipoproteinemic patients. Ann. N.Y. Acad. Sci. *393:* 222–223 (1982).

Heikkila, R.E.; Mezick, J.A.; Cornwell, D.G.: Destruction of specific membrane phospholipids during peroxidative hemolysis of vitamin E-deficient erythrocytes. Physiol. Chem. Phys. *3:* 93–97 (1971).

Horwitt, M.K.: Interrelations between vitamin E and polyunsaturated fatty acids in adult men. Vitams Horm. *20:* 541–558 (1962).

Horwitt, M.K.; Century, B.; Zeman, A.A.: Erythrocyte survival time and reticulocyte levels after tocopherol depletion in man. Am. J. clin. Nutr. *12:* 99–106 (1963).

Horwitt, M.K.; Harvey, C.C.; Dahm, C.H., Jr.; Searcy, M.T.: Relationship between tocopherol and serum lipid levels for determination of nutritional adequacy. Ann. N.Y. Acad. Sci. *203:* 223–236 (1972).

Hussain, Q.Z.; Newcomb, T.F.: Thrombin stimulation of platelet oxygen consumption rate. J. appl. Physiol. *19:* 297–300 (1964).

Jager, F.C.: Linoleic acid intake and vitamin E requirement in rats and ducklings. Ann. N.Y. Acad. Sci. *203:* 199–211 (1972).

Johnson, G.J.; Finkel, B.; Vatassery, G.; White, J.G.; Allen, D.W.: Shortened erythrocyte survival in chronic hemolytic disease due to glucose-6-phosphate dehydrogenase deficiency is not corrected by high dose vitamin E therapy. Blood *58:* suppl. 29a (1981).

Johnson, G.J.; Vatassery, G.T.; Finkel, B.; Allen, D.W.: High-dose vitamin E does not decrease the rate of chronic hemolysis in glucose-6-phosphate dehydrogenase deficiency. New Engl. J. Med. *308:* 1014–1017 (1983).

Kater, R.M.H.; Unterecker, W.J.; Kim, C.Y.; Davidson, C.S.: Relationship of serum tocopherol to beta-lipoprotein concentrations in liver diseases. Am. J. clin. Nutr. *23:* 913–918 (1970).

Kayden, H.J.; Silber, R.: The role of vitamin E deficiency in the abnormal autohemolysis of acanthocytosis. Trans. Ass. am. Physns *78:* 334–342 (1965).

Kelleher, J.; Losowsky, M.S.: The absorption of α-tocopherol in man. Br. J. Nutr. *24:* 1033–1047 (1970).

Kitabchi, A.E.; Wimalasena, J.: Demonstration of specific binding sites for ^{3}H-RRR-alpha-tocopherol on human erythrocytes. Ann. N.Y. Acad. Sci. *393:* 300–314 (1982).

Kitabchi, A.E.; Wimalasena, J.: Specific binding sites for *D*-α-tocopherol on human erythrocytes. Biochim. biophys. Acta *684:* 200–206 (1982).

Klebanoff, S.J.: Myeloperoxidase-halide-hydrogen peroxide antibacteria system. J. Bact. *95:* 2131–2138 (1968).

Klebanoff, S.J.; Clark, R.H.: Myeloperoxidase-H_2O_2-halide system antimicrobial activity; in The neutrophil function and clinical disorder, pp. 410–434 (Elsevier, Amsterdam 1978).

Koppenol, W.H.; Butler, J.: Mechanism of reactions involving singlet oxygen and the superoxide anion. FEBS Lett. *83:* 1–6 (1977).

Lake, A.M.; Stuart, M.J.; Oski, F.A.: Vitamin E deficiency and enhanced platelet function: reversal following vitamin E supplementation. J. Pediat. *90:* 722–725 (1977).

Levander, O.A.; Morris, V.C.; Ferretti, R.J.: Comparative effects of selenium and vitamin E in lead-poisoned rats. J. Nutr. *107:* 378–382 (1977).

Lewis, L.A.; Quaife, M.L.; Page, I.H.: Lipoproteins of serum, carriers of tocopherol. Am. J. Physiol. *178:* 221–222 (1954).

Losowsky, M.S.; Kelleher, J.; Walker, B.E.; Davies, T.; Smith, C.L.: Intake and absorption of tocopherol. Ann. N.Y. Acad. Sci. *203:* 212–222 (1972).

Lubin, B.; Chiu, D.: Properties of vitamin E-deficient erythrocytes following peroxidant injury. Pediatrics, Springfield *16:* 928–932 (1982).

Lucy, J.A.: Functional and structural aspects of biological membranes: a suggested structural role of vitamin E in the control of membrane permeability and stability. Ann. N.Y. Acad. Sci. *203:* 4–11 (1972).

Luzzato, L.; Testa, U.: Human erythrocyte glucose-6-phosphate dehydrogenase: structure and function in normal and mutant subjects. Curr. Top. Hematol. *1:* 1–70 (1978).

Machlin, L.J.; Filipski, R.; Willis, A.L.; Kuhn, D.C.; Brin, M.: Influence of vitamin E on platelet aggregation and thrombocythemia in the rat. Proc. Soc. exp. Biol. Med. *149:* 275–277 (1975).

Machlin, L.J.; Gabriel, E.: Kinetics of tissue α-tocopherol uptake and depletion following administration of high levels of vitamin E. Ann. N.Y. Acad. Sci. *393:* 48–60 (1982).

McCay, P.B.; Pfeifer, P.M.; Stipe, W.H.: Vitamin E protection of membrane lipids during electron transport functions. Ann. N.Y. Acad. Sci. *203:* 62–73 (1972).

McCormick, E.C.; Cornwell, D.G.; Brown, J.B.: Studies on the distribution of tocopherol in human serum lipoproteins. J. Lipid Res. *1:* 221–228 (1960).

Melhorn, D.K.; Gross, S.; Lake, G.A.; Leu, J.A.: The hydrogen peroxide fragility test and serum tocopherol level in anemias of various etiologies. Blood *37:* 438–446 (1971).

Mino, D.B.: Oxygen poisoning and vitamin E deficiency. J. Nutr. Sci. Vitaminol. *19:* 95–104 (1973).

Mino, M.; Kijima, Y.; Nishida, Y.; Nakagawa, S.: Difference in plasma and red blood cell tocopherols in breast-fed and bottle-fed infants. J. Nutr. Sci. Vitaminol. *26:* 103–112 (1980).

Mino, M.; Kitagawa, M.; Nakagawa, S.: Changes of alpha-tocopherol levels in red blood cells and plasma with respect to hemolysis induced by dialuric acid in vitamin E-deficiency rats. J. Nutr. Sci. Vitaminol. *27:* 199–207 (1981).

Mino, M.; Nakagawa, S.; Tamai, H.; Miki, M.: Clinical evaluation of red blood cell tocopherol. Ann. N.Y. Acad. Sci. *393:* 175–178 (1982).

Mohler, D.N.; Majerus, P.W.; Minnich, V.; Hess, C.E.; Garrick, M.D.: Glutathione synthetase deficiency as a cause of hereditary hemolytic disease. New Engl. J. Med. *283:* 1253–1257 (1970).

Molenaar, I.; Hulstaert, C.E.; Vos, J.: Membrane characteristics in vitamin E deficiency and the assessment of vitamin E status. Proc. Nutr. Soc. *32:* 249–254 (1973).

Molenaar, I.; Vos, J.; Jager, F.C.; Hommes, F.A.: The influences of vitamin E deficiency on biological membranes. An ultrastructural study on the intestinal epithelial cells of ducklings. Nutr. Metab. *12:* 358–370 (1970).

Moyer, W.T.: Vitamin E levels in term and premature newborn infants. Pediatrics, Springfield *6:* 893–896 (1950).

Nakamura, T.; Hishinuma, I.: Protective effect of tocopherol in the formation of TBA reactive materials in rat liver; in DeDuve, Hayaishi, Tocopherol, oxygen and biomembranes, pp. 95–109 (Elsevier, Amsterdam 1978).

Niki, E.; Tsuchiya, J.; Tanimura, R.; Kamiya, Y.: Regeneration of vitamin E from α-chromanoxyl radical by glutathione and vitamin C. Chem. Lett. *1982:* 789–792 (1982).

Nitowsky, H.M.; Cornblath, M.; Gordon, H.H.: Studies of tocopherol deficiency in infants and children. II. Plasma tocopherol and erythrocyte hemolysis in hydrogen peroxide. J. Dis. Child. *92:* 164–174 (1956).

Nordoy, A.; Day, H.J.; Lund, S.: Subcellular localization of human platelet phospholipids and their fatty acid and aldehyde composition. Scand. J. clin. Lab. Invest. *23:* 169–176 (1969).

Nordoy, A.; Strom, E.: Tocopherol in human platelets. J. Lipid Res. *16:* 386–391 (1975).

O'Brien, R.T.; Pearson, H.A.: Physiologic anemia of the newborn infant. J. Pediat. *79:* 132–138 (1971).

Okuma, M.; Steiner, M.; Baldini, M.G.: Studies on lipid peroxides in platelets. II. Effect of aggregating agents and platelet antibody. J. Lab. clin. Med. *77:* 728–742 (1971).

Olson, R.E.: Creatine kinase and myofibrillar protein in hereditary muscular dystrophy and vitamin E deficiency. Am. J. clin. Nutr. *27:* 1117–1129 (1974).

Oski, F.A.; Barnes, L.A.: Vitamin E deficiency: a previously unrecognized cause of hemolytic anemia in premature infants. J. Pediat. *70:* 211–220 (1967).

Packer, J.E.; Slater, T.F.; Willson, R.L.: Direct observation of a free radical interaction between vitamin E and vitamin C. Nature, Lond. *278:* 737–738 (1979).

Parrish, D.B.: Determination of vitamin E in foods – a review. CRC crit. Rev. Food Sci. Nutr. *13:* 161–187 (1980).

Peake, I.R.; Bieri, J.G.: Alpha and gamma tocopherol in the rat: in vitro and in vivo tissue uptake and metabolism. J. Nutr. *101:* 1615–1622 (1971).

Peake, I.R.; Windmueller, H.G.; Bieri, J.G.: A comparison of the intestinal absorption, lymph and plasma transport, and tissue uptake of α- and γ-tocopherols in the rat. Biochim. biophys. Acta *260:* 679–688 (1972).

Pearson, C.K.; McBarnes, M.: The absorption and distribution of the naturally occurring tocochromanols in the rat. Br. J. Nutr. *24:* 581–587 (1970).

Pelkonen, R.: Plasma vitamins A and E in the study of lipid and lipoprotein metabolism in coronary heart disease. Acta med. scand. *174:* suppl. 399, pp. 64–69 (1963).

Pickett, W.C.; Cohen, P.: Mechanism of the thrombin-mediated burst in oxygen consumption by human platelets. J. biol. Chem. *251:* 2536–2538 (1976).

Rachmilewitz, E.A.; Kornberg, A.; Acker, M.: Vitamin E deficiency due to increased consumption in beta-thalassemia and in Gaucher's disease. Ann. N.Y. Acad. Sci. *393:* 336–347 (1982).

Rachmilewitz, E.A.; Shohet, S.B.; Lubin, B.H.: Lipid membrane peroxidation in beta-thalassemia major. Blood *47:* 495–505 (1976).

Recommended Dietary Allowance; 7th ed. (National Academy of Sciences, Washington 1965).

Recommended Dietary Allowance; 8th ed. (National Academy of Sciences, Washington 1974).

Recommended Dietary Allowance; 9th ed. (National Academy of Sciences, Washington 1980).

Ritchie, J.H.; Fish, M.B.; McMasters, V.; Grossman, M.: Edema and hemolytic anemia in premature infants. New Engl. J. Med. *279:* 1185–1190 (1968).

Rubenstein, H.M.; Dietz, A.A.; Srinavasan, R.: Relation of vitamin E and serum lipids. Clinica chim. Acta *23:* 1–6 (1969).

Sayare, M.; Fikiet, M.; Paulus, J.: Effect of vitamin E on the binding of hemoglobin to the red cell membrane. Ann. N.Y. Acad. Sci. *393:* 251–262 (1982).

Scott, M.L.: Studies on vitamin E and related factors in nutrition and metabolism; in DeLuca, Suttie, The fat-soluble vitamins, pp. 355–368 (University of Wisconsin Press, Madison 1969).

Scott, M.L.; Desai, I.D.: The relative anti-muscular dystrophy activity of the *d*- and *l*-epimers of alpha-tocopherol and of other tocopherols in the chick. J. Nutr. *83:* 39–43 (1964).

Shapiro, S.S.; Mott, D.J.: Alterations of enzymes in the red blood cell membrane in vitamin E deficiency. Ann. N.Y. Acad. Sci. *393:* 263–276 (1982).

Silber, R.; Winter, R.; Kayden, H.J.: Tocopherol transport in the rat erythrocyte. J. clin. Invest. *48:* 2089–2095 (1969).

Simon, E.J.; Eisengart, A.; Sundheim, L.; Milhorat, A.T.: The metabolism of vitamin E. II. Purification and characterization of urinary metabolites of α-tocopherol. J. biol. Chem. *221:* 807–817 (1956).

Soderhjelm, P.; Anderson, B.: Simultaneous determination of vitamins A and E in feeds and foods by reversed phase high-pressure liquid chromatography. J. Sci. Fd Agric. *29:* 697–702 (1978).

Spielberg, S.P.; Garrick, M.D.; Corash, L.M.; Butler, J.D.; Tietze, F.; Rogers, L.; Schulman, J.D.: Biochemical heterogeneity in glutathione synthetase deficiency. J. clin. Invest. *61:* 1417–1420 (1978).

Steiner, M.: Vitamin E changes the membrane fluidity of human platelet. Biochim. biophys. Acta *640:* 100–105 (1981).

Steiner, M.: Effect of alpha-tocopherol administration on platelet function in man. Thromb. Haemostasis *49:* 1–5 (1983).

Steiner, M.; Anastasi, J.: Vitamin E. An inhibitor of the platelet release reaction. J. clin. Invest. *57:* 732–737 (1976).

Steiner, M.; Mower, R.: Mechanism of action of vitamin E on platelet function. Ann. N.Y. Acad. Sci. *393:* 289–299 (1982).

Stocks, J.; Dormandy, T.L.: The autoxidation of human red cell lipids induced by hydrogen peroxide. Br. J. Haemat. *20:* 95–111 (1971).

Strauss, R.R.; Paul, B.B.; Jacobs, A.A.; Sbarra, A.J.: Role of the phagocyte in host-parasite interactions. XXII. H_2O_2-dependent decarboxylation and deamination by myeloperoxidase and its relation to antimicrobial activity. J. reticuloendoth. Soc. *7:* 754–761 (1970).

Stuart, M.J.: Vitamin E deficiency: Its effect on platelet vascular interaction in various pathologic states. Ann. N.Y. Acad. Sci. *393:* 277–288 (1982).

Stuart, M.J.; Oski, F.A.: Vitamin E and platelet function. Am. J. Pediat. Hematol. Oncol. *1:* 77–82 (1979).

Svanholm, U.; Bechgaard, K.; Parker, V.D.: Electrochemistry in media of intermediate acidity. 8. Reversible oxidation products of the alpha-tocopherol model compound. Cation radical, cation and dication. J. Am. chem. Soc. *96:* 2409–2413 (1974).

Takahashi, Y.; Uruno, K.; Kimura, S.: Vitamin E binding proteins in human serum. J. Nutr. Sci. Vitaminol. *23:* 201–209 (1977).

Tangney, C.C.; McNair, H.M.; Driskell, J.A.: Quantitation of individual tocopherols in plasma, platelets, lipids, and livers by high performance liquid chromatography. J. Chromatogr., biomed. Appl. *224:* 389–397 (1981)

Tappel, A.L.: Vitamin E as the biological lipid antioxidant. Vitams Horm. *20:* 493–510 (1962).

Vatassery, G.T.; Krezowski, A.M.; Eckfeldt, J.H.: Vitamin E concentrations in human blood plasma and platelets. Am. J. clin. Nutr. *37:* 1020–1024 (1983).

Wever, R.; Oudega, B.; VanGelder, B.F.: Generation of superoxide radical during the autooxidation of oxyhemoglobin. Biochim. biophys. Acta *302:* 475–478 (1973).

Weber, F.; Wiss, O.: Über den Stoffwechsel des Vitamin E in der Ratte. Helv. physiol. pharmacol. Acta *21:* 131–141 (1963).

Welsh, S.O.: The protective effect of vitamin E and N,N'-diphenyl-p-phenylenediamine (DPPD) against methyl mercury toxicity in the rat. J. Nutr. *109:* 1673–1681 (1979).

Witting, L.A.; Lee, L.: Dietary levels of vitamin E and polyunsaturated fatty acids and plasma vitamin E. Am. J. clin. Nutr. *28:* 571–576 (1975).

Wright, S.W.; Filer, L.J., Jr.; Mason, K.E.: Vitamin E blood levels in premature and full-term infants. Pediatrics, Springfield *7:* 386–393 (1951).

Zalkin, H.; Tappel, A.L.; Caldwell, K.A.; Shibko, S.; Desai, I.D.; Holliday, T.A.: Increased lysomal enzymes in muscular dystrophy of vitamin E-deficient rabbits. J. biol. Chem. *237:* 2678–2682 (1962).

Ching K. Chow, PhD, Department of Nutrition and Food Science,
University of Kentucky, 212 Funkhouser Building, Lexington, KY 40506 (USA)

Wld Rev. Nutr. Diet., vol. 45, pp. 167–197 (Karger, Basel 1985)

Nutritional and Hormonal Requirements of Mammalian Cells in Culture

David Barnes

Department of Biological Sciences, University of Pittsburgh, Pittsburgh, Pa., USA

Contents

Introduction

Although experimental protocols involving the use of intact animals have been of tremendous value over past decades for studies of nutrition, physiology and endocrinology, these in vivo experiments are limited in many aspects by the potential complications of multiple interactions that take place among cells and tissues in vivo. The advantages of the use of in vitro cell culture systems for studies of these types have been recognized for many years, but experiments using cultured animal cells also have been limited somewhat by the difficulty in developing chemically and hormonally defined culture media for the maintenance and growth of animal cells in vitro. Conventional cell culture techniques have relied in general on the use of biological fluids or extracts such as serum, plasma, lymph, milk or embryo extract as a source of unknown or poorly defined components necessary for cell survival and growth outside the body, and the uncontrolled, inconsistent and undefined nature of these media supplements made precise studies of nutritionally and hormonally related phenomena difficult or impossible. Several innovative approaches have been taken in recent years toward the development of more defined culture media for mammalian cells in vitro and particularly rapid progress has been made in the last decade in this area.

Early attempts in use of defined culture media depended primarily on 'adaptation' of cells to a predetermined serum-free medium formulation [60]. This procedure in most cases selected for a small subpopulation within the population of cells used to initiate the cultures. Selected subpopulations of this type are often of limited value for studies of the nutritional and endocrinological aspects of normal cell physiology because the cells are variant from the original population with regard to their requirements and abilities to synthesize or function in the absence of some nutritional or hormonal components. Studies of these types, then, are in general carried out with cell populations that may not accurately represent the properties of the cells of the tissues from which they were derived originally and, furthermore, usually have been carried out with established cell lines derived by culture in serum-containing media of the parent population from which the variants were selected. These parent lines may be themselves already changed in many respects from the cells of the original tissue from which they were derived. Little success has been achieved at establishing cultures directly from primary tissue sources in vivo by 'adaptation' to serum-free media.

*Serum-Free Animal Cell Culture for Studies of
Nutrition and Endocrinology*

More recent approaches to the development of defined, serum-free culture conditions have been pioneered primarily by the laboratories of Dr. *Richard Ham* of the University of Colorado [52–56, 86, 99, 136, 140] and Dr. *Gordon Sato* of the University of California [11, 12, 14, 18, 21, 58, 81, 106, 110, 116, 117]. These approaches have allowed the development of methods for the maintenance and growth of the entire cell population to be studied, rather than only variant subpopulations derived by selective growth in vitro. Because the approaches were designed to be applicable to the bulk of the cell population in the cultures studied, the methods developed have also found application for primary cell culture [2, 8, 9, 11, 12, 14, 16, 23, 31, 35, 43, 54, 71, 78, 81, 98, 99, 112, 132, 136, 138].

The laboratory of Dr. *Ham* has approached the development of defined cell culture conditions by making improvements in the basal nutrient culture media that allow the progressive reduction or elimination of undefined (e.g. serum) supplementation of the medium [52–56, 86, 99, 136]. This approach primarily has been applied to clonal growth of a number of cell types, but the basal nutrient media formulations developed in this way are also applicable to conditions of mass culture in some cases [54, 56, 58, 81, 123, 145]. The approach has been successful at least partially because many of the protein components of serum that are active at promoting growth of cells in culture may do so by mechanisms related to regulation of the availability of nutrients to the cells [11, 12, 54, 56, 85]. Included in this class of growth-stimulatory serum factors are metal-binding proteins such as transferrin and ceruloplasmin [58, 80–82, 100, 101, 135], lipid carriers such as lipoproteins [49, 50, 57], and albumin [4, 64, 65, 114, 143, 147], and other binding proteins that may act to transport or stabilize nutritional components such as some vitamins that are chemically unstable or poorly soluble in aqueous media. Serum also, of course, contributes directly major nutritional components to culture media to which it may be added as a supplement [53–56]. This approach toward eliminating or reducing serum in culture medium also has been successful partially because complicated interactions exist between the availability of some nutritional components and some hormonal components supplied by serum [54, 76, 85]. In fact, it was this approach to the development of more defined culture media that pointed in a general way to the realization that the growth requirements of cells for some hormones can be replaced by a properly balanced nutritional environment [52–56, 85].

The laboratory of Dr. *Sato* has approached the development of more defined cell culture media by developing methods for the direct replacement of components of serum that are active in supporting the maintenance and growth of cells in culture with purified components capable of replacing the functions carried out by serum [11, 12, 14, 16, 21, 58, 106, 110, 116, 117]. These factors may be divided into the following groups: hormones, supplementary nutrients, binding proteins that modulate the action of hormones and nutrients, attachment factors and extracellular matrix components and extracellular enzymes [8, 9, 11, 12, 14]. Some of the components of these groups may be purified directly from serum. These include the iron-binding protein transferrin and the copper-binding protein ceruloplasmin [58, 100, 101, 135], albumin used primarily in serum-free media of this type as a lipid-binding protein [95, 114, 121, 122, 143, 146, 147], lipoprotein particles such as low-density lipoprotein (LDL) and high-density lipoprotein (HDL) [29, 57], substratum components fibronectin [97] and serum spreading factor [15–17], and the growth-stimulatory enzyme thrombin [33, 47, 119]. Many of the hormones or hormone-like growth factors used in serum-free media formulations of this type also are found in serum at concentrations that suggest they contribute to the overall growth-stimulatory capacity of the serum supplement to conventional culture media. These include insulin and insulin-like peptides or somatomedins, glucocorticoids, sex steroids, triiodothyronine, glucagon, vasopressin, platelet-derived growth factor, epidermal growth factor and fibroblast growth factor [11, 12]. Some of the hormonal supplements to these serum-free media, however, are probably not found in serum at concentrations high enough to expect that they might contribute to effects of the serum supplement in culture media. An example of this type of factors are the hormone-like hypothalamic releasing factors which may reach relatively high local concentrations in brain and are growth-stimulatory for some cell types in serum-free media but not generally found at mitogenic concentrations in the circulation. The observations that substances of this type affect cells in culture suggests that the responses may be indicative of physiological effects of these or similar factors in vivo, even in cases in which the responsive cells are not those conventionally considered to be the target cell type for the molecule [11, 12, 116].

For many cell types in culture, the greatest advantages in the development of defined or serum-free media lie in a combination of approaches such as the use of a basal nutrient medium of the type developed by the methods of *Ham* and co-workers [53–56] designed for growth of cells with little or no serum, supplemented with specific combinations of hormones, binding proteins, attachment factors [8, 9, 11, 12, 14] and, in some cases, extracellular enzymes such as thrombin or catalase [33, 38, 47, 119]. Combinations of these two approaches have led to considerable insights into the biology of animal cells in culture with regard to both the nature of the environmental factors affecting cell growth, metabolism and differentiation and the mechanisms by which the effects occur [11, 12, 54, 55, 76, 85, 145].

It is the purpose of this review to examine in some detail the information that experiments using the approaches described for the growth of mammalian cells in culture in serum-free or defined media have provided regarding the nutritional and hormonal requirements of these cells in vitro.

Nutritional Requirements of Mammalian Cells in Culture: Recent Insights

Metals and Trace Elements

The effects of several classes of nutritional factors on mammalian cells in culture are masked by the presence of serum in the culture medium. Among these are the trace elements – nutritional components that can be demonstrated with difficulty in whole animal experiments to be important to varying degrees for normal growth and development and maintenance of the healthy adult state. In particular, selenium has been shown to be important for long-term growth of several cell types in culture, including human fibroblasts [57, 86]. Other cell types that respond to a considerable degree to the inclusion in serum-free media of selenious acid or sodium selenite at concentrations of 1–10 nM are neuroblastoma, myeloma or hybridoma, embryonal carcinoma, several types of human lung carcinoma and canine and monkey kidney cells [11, 12, 18, 21, 30, 89, 94, 111, 128, 131]. Although the mechanism by which selenium stimulates cell growth in serum-free culture media has not been studied in great detail, it is assumed that in many of these circumstances the trace element, functioning as a cofactor for glutathione peroxidase, is involved in protection of cells from adverse effects of hydrogen peroxide or other peroxides generated in the cells or the media [51, 96, 141]. In some cases selenium in the medium can be replaced by similarly functioning factors such as catalase [38].

Although requirements of some other trace elements such as zinc and cadmium for optimal growth have been demonstrated in some cell culture systems, effects of these factors are somewhat difficult to establish, even in serum-free media, because they are effective at nanomolar or picomolar concentrations and may be commonly present as contaminants of other medium components, particularly the inorganic salts present at relatively high concentrations in basal nutrient media [53–56]. A mixture of trace elements including manganese, molybdenum, nickel, selenium, silicon, tin, vanadium and cadmium at concentrations in the range of 0.5–250 nM has been used in some serum-free systems, although absolute effects of all of these components have not yet been demonstrated in vitro [53–57, 61, 62, 86]. Caution must be exercised in the use of this mixture for some cell types in culture because some of these elements (e.g. manganese) may be toxic for some cell types, and other modifications of the composition of the mixture may be necessary to demonstrate stimulatory effects of some of the components [19–21, 58, 59].

Iron is an important and essential component of serum-free culture media, although effects of iron also may be difficult to demonstrate because of residual iron present as a contaminant of media components. Marked stimulation of cell growth by the addition of transferrin to serum-free, hormone-supplemented media probably represents the activity of this protein as a carrier and transporter of iron into cells, although other possibilities for growth-stimulatory mechanisms of transferrin also exist, such as detoxification of media by binding to transferrin of toxic heavy metals like lead [11, 12, 51, 58, 64, 65, 100, 135]. The work from *Ham's* laboratory indicates that in some cases the iron-related requirement of cells for transferrin can be reduced or eliminated completely by the addition to the medium of ferrous iron [53–56]. While ferrous iron is a component of some basal nutrient media, such as F12, iron in ferric form is often the form recommended for other media, such as Eagle's medium or Dulbecco-modified Eagle's medium [52–56]. These media were developed for use with a serum supplement, and probably rely on transferrin supplied by the serum for transport of ferric iron into the cells. Serum albumin is also capable of binding metal ions and possibly may act in culture medium as an iron carrier protein [4, 64, 65, 117].

Copper is important for growth of at least some cell types in vitro, and the copper carrier protein ceruloplasmin has been shown to be growth stimulatory for cells in serum-free culture. For both transferrin and ceruloplasmin, the concentration of protein necessary for growth stimulatory effects, in the range of 1–100 μg/ml culture medium, is not unreasonable when one considers the concentrations of these proteins in plasma or serum (for example, about 3 mg/ml for transferrin). The possibility exists for preparations of both transferrin and ceruloplasmin, however, that growth-stimulatory effects observed for these proteins in serum-free culture are due in at least some cases to contaminants of the preparations used [11, 12]. Both of these components are isolated from serum, and contamination by serum growth factors at a level of even one part in 10,000 might be high enough to stimulate cell growth. Another factor that is found in some sera and is stimulatory for the growth of some cell types in serum-free medium is the tripeptide glycyl-histidyl-lysine [2, 5]. This factor binds both copper and iron and enhances the uptake of copper into hepatoma cells in culture [102].

Experiments in both serum-free and serum-containing media have implicated calcium as an important nutritional regulator of cell growth, and interrelationships between mitotic responses to growth factors and calcium concentration in culture media have been documented. Optimal calcium concentration is changed in transformed cells, and changes in extracellular calcium concentration have also been shown to affect the differentiation potential of cells in culture [54, 85, 99, 136]. Although no specific calcium-binding protein probably is needed in cell culture in order to allow the utilization of calcium by the cells, a number of serum proteins could act in this way, and calcium also may interact with some of the lower molecular weight components of culture media under normal culture conditions.

Vitamins

Effects of a number of vitamins may be demonstrated in cell culture, and the use of serum-free culture media for the elucidation of mechanisms of vitamin action has the potential to provide important insights that would be difficult to reach in experiments utilizing whole animals. Ascorbic acid in serum-free media is growth-stimulatory for several cell types, and in some cases it is possible to demonstrate clearly that this effect is the result of increased collagen production in cultures containing ascorbic acid [92, 120]. Vitamin C is a cofactor for proline hydroxylase activities critical for collagen formation. Ascorbic acid is less important for cell survival and growth in culture if the appropriate collagen type is supplied from exogenous sources. In addition to its function as a specific enzymatic cofactor for hydroxylases, vitamin C also functions in culture medium as general antioxidant.

Another vitamin that acts as an antioxidant in culture medium is α-tocopherol, vitamin E. In vivo vitamin E deficiency results in male sterility in several species, and has been shown to be required for survival of porcine testicular Leydig cells in culture at concentrations in the range of 10–200 ng/ml [74–79, 82]. Ascorbic acid also was found to act in a similar manner as that of vitamin E, but considerably higher concentrations in the culture medium were required. Pig Leydig cells in serum-free culture also were shown to be vitamin E-responsive for chorionic gonadotrophin-stimulated testosterone secretion, chorionic gonadotrophin receptor maintenance and prostaglandin secretion, although it is unclear how these effects relate to the observed requirement of vitamin E for survival of these cells in vitro.

Several types of effects of vitamin A or derivatives of this vitamin have been observed in culture. Improved growth of testicular cells of Sertoli cell origin is seen in serum-free culture in the presence of retinol [74, 75, 81, 82]. Examples also exist of effects of retinol or retinoic acid on cellular differentiative potential both in vivo and in vitro in conventional serum-containing media or in serum-free media, and several lines of embryonal carcinoma and lymphoid cells that respond to compounds of this type in serum-free media show promise for studies at the cellular level of the molecular processes involved in triggering differentiation [26, 27, 66, 108, 109, 125, 126].

Lipids

In addition to studies of lipid-soluble vitamins like vitamins A and E, culture of cells in the absence of serum supplementation also has allowed the identification in a more precise way of other lipid requirements for growth of some cell types. Even essential lipids may prove toxic for cultured cells if presented in a form that exposes cells to the detergent action of lipids, and several different methods have been devised for providing necessary lipids in culture media. Cells may be provided with lipid in culture by exposure to artificially constructed liposomes containing various combinations of soybean lecithin, cholesterol, sphingomyelin, vitamin E, vitamin E acetate, phosphatidylcholine, phosphatidylethanolamine, phosphatidylinositol, and dipalmitoyl phosphatidic acid [53–56]. Cells also may be provided with fatty acids such as linoleic acid or oleic acid in a form bound to serum albumin [4, 51, 61, 64, 65, 95, 112, 114, 121, 143, 147]. In general, cells in culture in serum-free media have not been found to require for growth other saturated or unsaturated fatty acids such as palmitic, stearic or arachidonic. In addition to using liposomes and albumin as a carrier some success has also been reported at providing linoleic acid to cells in the form of dilinoleoyl phosphatidylcholine [37–39]. Another approach has been to use α-cyclodextrin in place of albumin [146].

Fulfillment of the lipid requirement of cells in culture also can be met by direct addition of plasma lipoprotein particles. HDL, providing glycerides, phospholipids (mostly phosphatidylcholine), cholesterol and cholesteryl esters, and LDL, providing cholesteryl ester, cholesterol, triglyceride and phospholipid in different relative proportions than that found in HDL, have been used for these purposes [29, 49, 50, 57, 134]. Although it is likely that many of the effects of HDL and LDL observed on cultured cells are due to the lipid portions of the particles, some evidence exists that the apoproteins of plasma lipoprotein particles may affect cells in the absence of extractable lipid [83].

Another lipid-related nutritional component that is growth-stimulatory for some cell types in culture in the absence of serum is ethanolamine or phosphoethanolamine [5–7, 67–69, 71, 94, 99, 123, 136]. The ethanolamine analogues monomethylethanolamine, 1-amino-2-propanol and 2-amino-1-propanol are also active in a manner similar to ethanolamine or phosphoethanolamine, but only at higher concentrations. Phosphoethanolamine is active at stimulating growth of the 64-24 rat mammary carcinoma in vitro at concentrations in the range of 1–5 μM [67–69]. Ethanolamine added exogenously to the culture medium is taken up by cells and incorporated into phospholipid (e.g. phosphatidylethanolamine), and it has been suggested that exogenous ethanolamine is stimulatory for cell growth because in its absence the cells are limited in growth rate by the amount of phosphatidylethanolamine that can be synthesized utilizing endogenous precursors [68]. Marked changes in phospholipid composition of responsive cell types occur after exposure to ethanolamine in the culture medium. Cell types responsive to ethanolamine in serum-free media in addition to rat mammary carcinoma include mouse hybridoma [94] and human epidermoid carcinoma, epidermal keratinocytes, bronchial epithelium and mammary epithelial cells [5–7, 71, 99, 123, 136].

Polyamines

Putrescine is another nutritional component that is found to be stimulatory for the growth or survival of some cell types as the serum supplement to the culture medium is lowered or eliminated [18–23, 44, 73, 90, 91, 148]. Polyamines and polyamine synthesis are recognized as being intimately involved in the processes of rapidly dividing cells, and cell types such as neuroblastoma that require high concentrations (e.g. 100 μM) of exogenously added putrescine for growth may exhibit this requirement because of deficiencies in putrescine synthesis or transport that render the concentration of this nutrient in the cell rate-limiting for growth [77].

Hormonal Requirements of Mammalian Cells in Culture

Serum-free cell culture techniques have proved to be very useful in the identification and study of hormonal requirements for the maintenance and growth of cells in vitro, and also have helped identify in some cases those factors that affect the differentiative capacity of some types of cultured mammalian cells [9, 11, 12, 14, 16, 21, 106, 110, 117]. In this section will be reviewed some of the information derived by the techniques of serum-free cell culture regarding the hormonal responses of a number of cell types.

Cells of the Endocrine System

The first cell type to be studied in detail in serum-free medium by the laboratory of *Sato* was the GH3 rat pituitary carcinoma cell line [58, 59]. This line was established in culture by *Yasumura* et al. [149] from a transplantable rat pituitary tumor line [127], and expresses in vivo and in vitro a number of differentiated functions associated with normal pituitary cells, including secretion of prolactin and growth hormone. When added to Ham's F12 as a basal nutrient medium, growth-stimulatory activity is seen for triiodothyronine (from thyroid), thyrotropin-releasing hormone (from hypothalamus), parathyroid hormone (from parathyroid), insulin (from pancreas), fibroblast growth factor (from pituitary) and somatomedin C (isolated from plasma and probably synthesized by the liver) [58, 59, 144, 145]. All of these factors are active at promoting GH3 cell growth in serum-free culture at concentrations that one might consider physiologically relevant (e.g. nanograms per milliliter) except for insulin, which is growth-stimulatory only at supraphysiological concentrations (micrograms per milliliter). The reasons that insulin is required at high concentrations are probably related to rapid inactivation of the hormone in serum-free culture media in the absence of stabilizing proteins, and also may reflect the ability of insulin at high concentrations to mimic the effects of insulin-like growth factors found in plasma or serum [59].

In addition to the information serum-free cell culture has provided regarding the nutritional requirements of cells of the testes, significant information has come from these types of studies regarding the hormonal requirements of several of the cell types of this organ. Sertoli cells require for optimal growth in serum-free medium insulin, epidermal growth factor, follicle-stimulating hormone, somatomedin C and growth hormone [74, 75, 81, 82]. While the growth-stimulatory capacity of follicle-stimulating hormone and growth hormone on these cells is unusual when compared to the hormonal requirements of most cell types in serum-free culture, the requirements for insulin and epidermal growth factor reflect a broad target specificity for these factors. Virtually every cell type examined in serum-free culture by the methods of *Sato* and co-workers has been found to require insulin for optimal growth, although this hormone is more critical for survival and growth for some cell types than for others [11, 12, 59]. The target cell specificity of epidermal growth factor, while not as sweeping as insulin, is still quite broad. Epidermal growth factor acts at very low concentrations to stimulate growth, and functions through high-affinity plasma membrane receptors. For the purposes of this review this hormone-like peptide will be considered a hormone in the same sense as insulin or trophic pituitary peptides are considered hormones, although epidermal growth factor does not fit the definition of a hormone in a strict, classical sense.

Another cell type of the endocrine system for which serum-free hormonal requirements have been identified are ovarian cells. RF1 rat ovarian cells can be maintained in long-term culture, in a fairly simple serum-free medium, consisting of a one-to-one mixture of Ham's F12 and Dulbecco-modified Eagle's medium supplemented with insulin, transferrin, hydrocortisone and attachment factors fibronectin or serum-spreading factor [15, 97, 98]. In serum-free medium primary rat ovarian cells can be demonstrated to maintain biochemical responsiveness to follicle-stimulating hormone with regard to steroidogenesis (production of progestins and estrogens), expression of functional luteinizing hormone receptors and morphogenetic changes [97, 98]. These effects of follicle-stimulating hormone in vitro are generally much more difficult to demonstrate in conventional, serum-containing media, apparently due to inhibitory or blocking effects of components of the serum supplement.

Another cell type of the endocrine system for which hormonal responses to trophic hormones can be demonstrated in serum-free culture are rat thyroid cells. Cultures of these cells in an appropriate culture medium supplemented with insulin, transferrin, hydrocortisone, somatostatin and glycyl-histidyl-lysine have been shown to require thyrotropin for survival and growth, and to express thyroid-specific functions in vitro including iodide concentration and thyroglobulin production [2].

Responses to a trophic hormone also can be demonstrated for bovine adrenocortical cells in serum-free culture [47, 49, 119]. In an appropriate medium supplemented with a number of hormonal and nutritional components as well as the extracellular enzyme thrombin, these cells can be shown to exhibit an inducible steroidogenic pathway that is stimulated by adrenocorticotropin. Bovine adrenocortical cells are sensitive in this response to several of the nutritional components of serum-free cell culture medium mentioned above, including LDL, ascorbic acid, α-tocopherol and selenium.

Mammary epithelia from rodent, as well as human, sources have been studied in detail in serum-free medium. The initial observation regarding the growth-stimulatory activity of phosphoethanolamine and ethanolamine in cells in culture [67, 68] were made with a rat mammary tumor cell line [69, 127]. This line in serum-free medium is also stimulated by insulin, hydrocortisone, triiodothyronine, estradiol, prolactin and epidermal growth factor. These cells also exhibit a mitogenic response in serum-free culture to cholera toxin, which increases adenosine 3′,5′-cyclic phosphate (cAMP) production by the cells through activation of adenyl cyclase at the cell membrane. Presumably the increase of intracellular cAMP produced by cholera toxin is mimicking the activity of some hormones that are growth-stimulatory for these cells.

Normal or tumorigenic mouse epithelia in serum-free culture in collagen gels also are stimulated by cholera toxin, insulin and epidermal growth factor [63, 148]. These cells, like rodent neuroblastoma, require increased amounts of exogenous putrescine in the medium. Normal rat mammary epithelia in monolayer culture in a basal nutrient medium consisting of improved Eagle's medium can be stimulated to grow by the addition of insulin, dexamethasone, epidermal growth factor and other components, including ascorbic acid, which is acting to stimulate collagen production for these cells [120]. Several other of these factors also act to increase the net collagen production by the cells. These include epidermal growth factor, which stimulates collagen synthesis, and dexamethasone, which inhibits collagen breakdown in the extracellular matrix by inhibiting the production of an active collagenase by the cells. Somatomedin C, dibutyryl cAMP and prostaglandin E_1 are also mitogenic for the cells in this culture system, but trophic hormones prolactin and estradiol are not mitogenic.

The MCF7 or ZR-75-1 human mammary carcinoma lines have been shown to be stimulated to grow in serum-free media by insulin, triiodothyronine, estradiol, prostaglandin, fibroblast growth factor and epidermal growth factor [1, 10, 13]. Dexamethasone also improves plating efficiency of ZR-75-1 [1], although dexamethasone or hydrocortisone are inhibitory for the growth of both ZR-75-1 and MCF7 in long-term, mass culture. Normal human mammary epithelia [124] also recently have been cultured in serum-free media, and found to be stimulated by insulin, hydrocortisone, epidermal growth factor and an unidentified component in bovine pituitary extract [123]. Although nutritional components found in pituitary extracts such as phosphoethanolamine stimulate cell growth [67], and ethanolamine or phosphoethanolamine are stimulatory for human mammary epithelia in vitro [124], the low level of pituitary extract required for stimulation of human mammary epithelial growth in vitro suggests that the factor is probably a hormone active at low levels. The basal nutrient medium in which the hormones listed above were found to be mitogenic for human mammary epithelia was specifically designed for the optimal serum-free growth of this cell type, and has been designated 'MCDB170' [124].

Normal or benign human prostatic epithelial cells have been grown in a different basal nutrient medium, RPMI 1640, supplemented with insulin and dexamethasone [32]. In addition to these hormones, several nutritional supplements must be added to the basal nutrient medium for optimal growth of these cells. Transferrin is added, presumably acting as an iron carrier, and the medium is also supplemented with zinc chloride.

Epithelial Cells

Other human carcinoma cell types for which hormonal and nutritional requirements have been examined in vitro include HeLa human cervical carcinoma [62, 144, 145]. The serum-free medium devised for the optimal growth of these cells employs Ham's F12 [52] as the basal nutrient formulation, supplemented with a mixture of trace elements devised by *Ham,* transferrin, and the hormonal supplements insulin, epidermal growth factor, fibroblast growth factor and hydrocortisone. The most important hormonal supplements for survival and growth of HeLa cells in serum-free culture are hydrocortisone and epidermal growth factor. Hydrocortisone can be replaced in this medium by aldosterone [62]. In MCDB105, an alternative basal nutrient medium also developed in *Ham's* laboratory, several of the hormonal supplements are no longer required for serum-free growth of HeLa cells [145]. The MCDB105 medium was originally developed for the growth of human fibroblasts [53, 55, 56, 140].

In experiments examining the growth in vitro of human colon carcinoma cells derived from transplantable tumor lines established and carried in nude athymic mice, it was observed that insulin, epidermal growth factor, hydrocortisone, gastrin and glucagon are mitogenic for these cells [92, 137, 138]. The growth of the cells is also improved by addition to the medium of ascorbic acid or by plating the cells on a preformed collagen gel. The stimulatory activity of gastrin and glucagon on these cells in serum-free media is somewhat unusual. Although glucagon has been observed to be stimulatory, along with insulin and epidermal growth factor, for kidney cells and hepatocytes [70, 72, 128], as well as a few other cell types in culture, virtually no other cell type yet examined has been reported to exhibit a mitogenic response to gastrin in serum-free media [11, 12].

A considerable body of information on the hormonal responses of kidney epithelia in serum-free medium exists, much of it derived from the initial work of *Taub* et al. [128] who examined the growth requirements in serum-free media of the well-characterized MDCK canine kidney cell line. These cells are responsive to insulin, prostaglandin E_1, triiodothyronine, hydrocortisone, epidermal growth factor, fibroblast growth factor, glucagon and norepinephrine [128–133]. Some of these factors will replace others; for instance, the stimulation of growth observed by addition of prostaglandin E_1 (an adenyl cyclase activator), which results in increased intracellular cAMP, can be mimicked by glucagon, norepinephrine, or the direct addition of dibutyryl cAMP or isobutyl methylxanthine, an inhibitor of phosphodiesterase. Prostaglandin E_2 will also replace prostaglandin E_1. Variant MDCK cells that do not require prostaglandin E_1 for growth have been isolated and have been found to have higher levels of cAMP than the parent cells [130]. The MDCK line exhibits transport properties similar to those observed in cells of the collecting duct or distal tubule of the kidney, and these cells in monolayer culture will form domes or hemicysts, indicative of transporting epithelium in vitro. Both prostaglandin and hydrocortisone in the medium are important for the maintenance of this property in culture.

A cell line derived from porcine kidney, LLC-PK1, exhibits properties in vitro similar to those of proximal kidney tubule cells [34, 35, 115], and can be demonstrated to possess an active transport system for glucose. These cells are stimulated to grow in vitro in serum-free medium by insulin, hydrocortisone, triiodothyronine, and vasopressin [34, 115]. Although prostaglandin E_1 is a potent mitogen for the MDCK line, it is not mitogenic for the LLC-PK1 line. Vasopressin, but not prostaglandin E_1, apparently stimulates cAMP production in the LLC-PK1 cell, and the growth-stimulatory effect of vasopressin on these cells can be mimicked by dibutyryl cAMP or phosphodiesterase inhibitors. As with MDCK, a mitogenic effect of epidermal growth factor on LLC-PK1 cells also can be observed. The LLC-PK1 line also shows a nutritional requirement for cholesterol in serum-free culture. While many cell types are capable of synthesizing cholesterol endogenously in vitro, they may not do so at a rate necessary for optimal growth, and cholesterol supplied exogenously in the medium may be necessary to achieve rapid growth rates. The source of exogenous cholesterol in vitro in conventional, serum-containing media is lipoprotein particles.

The hormonal requirements for epidermoid cells originating from several areas of the body have been examined in vitro in serum-free culture.

Human bronchogenic epidermoid carcinoma cells established from human lung tumors and carried as transplantable tumor lines in athymic mice have been grown in culture in serum-free medium supplemented with insulin, glucagon and nutritional supplements selenium and retinoic acid [89]. The presence of selenium, added as selenious acid, is particularly important for the serum-free growth of these cells. Retinoic acid is optimally effective at concentrations below 10 nM, and higher concentrations are inhibitory for growth. Hormones found to inhibit growth of these cells include epidermal growth factor and hydrocortisone. Transferrin also is inhibitory for the growth of the cells, but the mechanism of this effect has not been studied. The action of retinoic acid in promoting cell growth in serum-free cell culture seems to be related to the ability of this factor to prevent keratinization and stratification of the cultures. This effect of retinoic acid also can be mimicked by triiodothyronine, retinol, retinal, retinylacetate, retinylpalmitate and thyroxine. Human small cell lung cancer also has been grown in culture in serum-free medium, and these cells were found to be stimulated by insulin, hydrocortisone and, surprisingly, estradiol [30, 31, 88, 118].

Normal human bronchial epithelial cells, when cultured serum-free using as a basal nutrient medium a modification of *Ham's* MCDB150 series developed for human epidermal keratinocytes [99, 136], are stimulated to growth by addition of epidermal growth factor, insulin, hydrocortisone and, as nutritional supplements, phosphoethanolamine, ethanolamine, and trace elements [71]. Extracts from pituitary and brain also are stimulatory for the growth of these cells, although the identity of the mitogenic components in these extracts has not been established. Cholera toxin, as a stimulator of adenyl cyclase, is mitogenic for epidermal keratinocytes, but is not stimulatory in this system for bronchial epithelial cells. Human epidermal keratinocytes in serum-free culture are responsive to hydrocortisone, epidermal growth factor, insulin and, marginally, to progesterone [99, 136]. Elimination of a requirement for transferrin for these cells can be accomplished by adjustment of the iron and zinc concentrations in the medium, and the tendency of these cells to undergo terminal differentiation and cease proliferation in vitro can be controlled by adjustment of the calcium concentration. Human epidermal keratinocytes, like several other types of epidermoid cells, are responsive to the nutritional supplements ethanolamine and phosphoethanolamine.

The A431 line of human epidermoid carcinoma cells, used by many laboratories interested in the effects of epidermal growth factor because of the large number of receptors for this hormone expressed by these cells in culture, is stimulated to grow in serum-free medium by several hormones, including insulin, somatostatin and parathyroid hormone [9–11]. Like others of this cell type A431 cells are responsive to ethanolamine. A mitogenic effect of glycyl-histidyl-lysine observed for these cells may reflect a nutritional requirement for trace metal transport, although this has not been explored with the A431 line. Epidermal growth factor in serum-free medium at concentrations that are stimulatory for growth of many cell types is inhibitory for the growth of A431 cells.

Neural Cells

Serum-free cell culture techniques also have been applied to the study of the hormonal and nutritional requirements of cells of neural tissue in culture. Rodent and human neuroblastoma have been grown in serum-free media [18–23], and it has been observed that these cells are stimulated in vitro by insulin and progesterone and by supplementary nutrients: putrescine at micromolar concentrations, and selenium at nanomolar concentrations. In particular, the presence of selenium is required for long-term survival of the culture for more than a few days in the absence of serum. Postmitotic neurons in primary culture can be maintained in serum-free media in the presence of a similar mixture of factors, but these cells also require nerve growth factor for survival in culture [23]. Rat pheochromocytoma cells in culture exhibit a similar set of responses to extracellular hormones and nutrients in serum-free cell culture, but also respond in a positive manner to epidermal growth factor [48]. This hormone, as well as a number of other factors, is inhibitory for the growth of neuroblastoma [18–20]. Other hormonal factors inhibitory for the growth of neuroblastoma or other neuronal cell lines in serum-free cell culture include epinephrine, estradiol, glucagon, norepinephrine, prostaglandin D_2, prostaglandin E_2, testosterone and triiodothyronine. Nutritional factors inhibiting neuroblastoma growth include adenosine at concentrations greater than 10 μM, retinoic acid at greater than 0.3 nM and selenium at greater than 0.3 μM.

Fetal mouse hypothalamic cells in culture have been reported to exhibit a mitogenic response to insulin, progesterone, and estradiol [44]. Triiodothyronine, while not growth-stimulatory for these cells, does increase neurite elongation and survival of the cultures [105]. Important nutritional supplements to the culture medium for these cells are arachidonic acid, docosahexaenoic acid, and selenium. Critical for the growth of these cells is the presence of transferrin, presumably representing a nutritional requirement for exogenously supplied iron.

Rat gliomas in serum-free cell culture are responsive to insulin and fibroblast growth factor and exhibit a nutritional requirement for relatively high concentrations of unsaturated fatty acids, such as linoleic acid, provided to the cells in a form complexed to albumin [143]. Oleic, arachidonic and palmitic acid will partially replace linoleic acid for these cells. Exhibiting responses somewhat similar to those of glioma cells in serum-free culture, astrocytes have been shown to grow in vitro in response to insulin and fibroblast growth factor, as well as hydrocortisone, prostaglandin $F_{2\alpha}$ and epidermal growth factor [90, 91]. As with neuroblastoma, putrescine is an important nutrient supplement for serum-free growth of these cells. The concentration of prostaglandin used in this system is somewhat high, and considering the stimulatory effect of arachidonic acid on other cell cultures of neural origin, may represent at least in part a nutritional contribution of this hormone-like lipid. The requirement of exogenous putrescine in serum-free cultures of several types of cells of neural tissue origin may indicate that these cells in vivo are under the influence of a hormonal factor or factors, as yet unidentified, that functions as a potent stimulator of endogenous polyamine synthesis or of putrescine transport into the cell. In the absence of this factor in vitro, high concentrations of exogenous putrescine must be provided for optimal growth of the cells.

Although not cells of nerve tissue, melanocytes derive embryologically from the neural crest and migrate to the skin and hair follicles during development, and resemble in some respects neural cells in their nutritional and hormonal requirements in culture. Mouse melanoma cells in culture are responsive to testosterone but, like some cells derived from neural tissue, melanoma are also responsive to progesterone, and testosterone can be partially replaced by progesterone in serum-free culture medium designed for the growth of melanoma [80]. Progesterone is not stimulatory in this system because of metabolic conversion to testosterone, and progesterone and testosterone also do not seem to be acting through a common metabolite. Melanoma cells in serum-free culture also exhibit a mitogenic response to nerve growth factor, as well as several hormones that one might not expect to be active on cells of this type: follicle-stimulating hormone from the pituitary and luteinizing hormone-releasing hormone from the hypothalamus. Melanoma cells in serum-free culture also require insulin for optimal growth. Normal melanocytes in culture are responsive to cholera toxin, presumably acting by increasing cAMP concentrations inside the cells, and the tumor promoter phorbol 12-myristate-13-acetate (PMA). Melanocytes are stimulated to synthesize melanin by addition of melanocyte-stimulating hormones or melanotropins (MSH); effects of MSH on these cells are regulated by intracellular cAMP concentration and MSH effects in culture can be mimicked by dibutyryl cAMP or isobutylmethylxanthine, an inhibitor of phosphodiesterase [139].

Embryonal Carcinoma Cells

Although effects of hormonal or nutritional components of culture media on the differentiative potential of some cell types in culture have been reported, such as the effects of MSH or retinoic acid mentioned previously, extensive work in this area remains to be done. One of the most attractive systems for experiments of this kind is the in vitro, serum-free culture of embryonal carcinoma cells that differentiate to form cell types of various kinds. A number of lines of mouse embryonal carcinoma have been studied in serum-free media [40–42, 108–113]. Several lines are responsive mitogenically to insulin and insulin-like growth factors, as well as epidermal growth factor, the epidermal growth factor-like sarcoma growth factor and fibroblast growth factor. Embryonal carcinoma cell lines also may require for optimal growth nutritionally related supplements in either mass culture or at clonal density: transferrin, lipoproteins, or unsaturated fatty acid complexed to albumin, selenious acid and ethanolamine. Retinoic acid has been shown to be a potent stimulator of differentiation in some of these culture systems and hexamethylene bisacetamide is also effective in some.

Fibroblastic Cells

Considerable emphasis in both conventional, serum-containing cell culture techniques and in recent studies using more defined, serum-free media has been placed on fibroblastic cell types in vitro. The emphasis on these cell types probably is the result of the ease with which such cells can be cultivated in vitro, allowing extensive use of fibroblastic cells in studies of hormonal and nutritional control of cell proliferation.

In a carefully formulated basal nutrient medium developed for human diploid fibroblasts that includes, among other changes from conventional culture methods, a reduction in the carbon dioxide concentration of the atmosphere in which the cells are incubated, it has been shown that these cells are responsive to insulin, epidermal growth factor, platelet-derived growth factor and hydrocortisone [53–56, 140]. Transferrin is not essential for delivery of iron to human fibroblasts under these conditions, and can be replaced by the use of freshly made ferrous sulfate. Other low molecular weight supplements required for optimal growth of human diploid fibroblasts under these conditions include phospholipids, cholesterol, sphingo-myelin and α-tocopherol, prostaglandins, phosphoenolpyruvate and reducing agents. Some of these components are likely to be acting through mechanisms related to improvement of the extracellular environment of the cells (e.g. as detoxifying agents or buffers) rather than intracellularly as nutritional factors. Other stimulators of human fibroblast cell growth in vitro in other basal nutrient formulations include triiodothyronine, thrombin and ascorbic acid [54, 142, 147], the latter presumably acting as a stimulator of endogenous collagen production by the cells. Supplementation of the basal nutrient medium developed for human diploid fibroblasts with insulin, fibroblast growth factor and lipids also allows the growth in the absence of serum of rabbit chondrocytes in culture [54].

Approaches similar to those taken for the study of human fibroblasts have led to a precisely formulated, hormonally supplemented basal nutrient medium for the growth of fibroblasts of other species in order to study effects of neoplastic transformation on hormonal responses of these cells [33, 83, 84, 114]. A simple serum-free medium [103] also has been developed for the mouse C3H 10T1/2 cell line [104, 107], originally developed in the laboratory of *Heidelberger,* and used in recent years in the study of environmental factors including neoplastic transformation. These cells can be maintained long-term at suboptimal growth rates in a serum-free medium formulation that includes insulin and epidermal growth factor as hormonal supplements, and does not require the addition of exogenous trace elements such as selenium or lipids other than low levels of linoleic acid [103]. It is likely that, under these circumstances, the cells are obtaining trace elements as contaminants of medium components or the reagents used to passage the cells (e.g. trypsin and trypsin-inhibitor solutions). Modifications of the serum-free medium developed for the 10T1/2 cells have been used to establish growth-restrictive, selective conditions for the isolation of viral transformants of these cells after infection with simian virus 40 [103].

Lymphoid Cells

Although much of the attention devoted to the study of the nutritional and hormonal requirements of animal cells in culture has been directed toward models of anchorage-dependent cells derived from solid tissues, some work has appeared on the serum-free growth in vitro and hormonal and nutritional requirements of lymphoid cells in culture. The laboratory of Dr. *Norman Iscove* in Basel carried out the pioneering work in this area [51, 64, 65], and in early studies established that lymphocyte and hemopoietic cell cultures were stimulated to grow by nutritionally related factors, such as transferrin and lipid in a form complexed to albumin. In general, growth stimulation by the peptide or steroid hormones or hormone-like growth factors of the kind commonly found to be stimulatory for cells grown in monolayer is rarely seen for lymphoid cells. However, insulin, stimulatory for virtually every cell type examined in the absence of serum, is growth-stimulatory for lymphoid cells. The work of *Iscove* and co-workers [51, 64, 65] also helped establish the general importance in serum-free culture of nutritional factors, such as trace elements like selenium, for the long-term growth of cells in the absence of serum. Other nutritional components of the basal culture medium found to be particularly important for lymphoid cells include cystine at concentrations higher than those used in most basal nutrient media, and β-mercaptoethanol or α-thioglycerol [28, 51, 64, 65]. The latter two factors are active at promoting growth when added in either the reduced or oxidized state.

Requirements of cells for lipids in culture can be satisfied by a mixture containing phosphatidylcholine, phosphatidylethanolamine, phosphatidyl-inositol, cholesterol and sterolglycosides, triglycerides, vitamin E, and free fatty acids [51, 64, 65]. Different differentiated subtypes of lymphocytes and hemopoietic cells vary in their requirements for some of the lipid components in the list above. For instance, terminal erythroid maturation is dependent on cholesterol at high concentrations, but high concentrations of cholesterol can be toxic for other subtypes within the hemopoietic and lymphoid population. This may be due to inhibition of hydroxymethylglutarate CoA reductase by cholesterol, resulting in the inhibition of synthesis of mevalonic acid. Such a block may be overcome by addition of exogenous mevalonic acid to culture medium, but does not reverse the cholesterol-induced toxicity in at least some cases. It has been suggested that the inhibitory effect of excess cholesterol on lymphocyte growth may be related to oxidized impurities present in the preparations used to supplement the medium [51, 64, 65]. Other factors that have been reported to stimulate growth of lymphoid cells under one or another set of conditions in serum-free media include testosterone, epidermal growth factor, prostaglandin E_1, triiodothyronine, and glucagon, although the effects of some of these factors are marginal [93, 94]. Lipid requirements in serum-free media have been met for some types of lymphoid or hybridoma cell lines by addition of less complicated mixtures than those referred to above. In some cases, albumin-fatty acid complexes, dilinoleoyl phosphatidylcholine, ethanolamine, or phosphatidylethanolamine and phosphatidylglycerol will suffice [37–39, 61, 93, 94, 122].

The hormonal and nutritional requirements for growth and induction of differentiation of several lines of human myeloid leukemia lines have been examined in serum-free media. These include the K562 [3, 36] line, established from a pleural effusion of a patient with chronic myelogenous leukemia in blast crisis, the KG10 KG-1 line, established from a patient with acute myelogenous leukemia, and the HL60 line, established from a patient with acute promyelocytic leukemia [24–27, 45, 46]. The HL60 line can be induced to differentiate into mature granulocyte-type cells with a number of low molecular weight factors, including dimethylsulfoxide, dimethylformamide, hypoxanthine, butyric acid and retinoic acid [24–27]. Retinal, retinol and retinyl acetate are without effect on these cells. The only hormonal supplement essential for the growth of these cells is insulin. The line exhibits, not surprisingly, a requirement for exogenously supplied lipid and transferrin as an iron-binding protein. The K562 line can be induced to synthesize hemoglobin by either hemin or butyric acid, and both HL60 and KG-1 are reported to exhibit monocyte or macrophage-like properties upon treatment in vitro with phorbol diester, this effect requiring the presence of neither transferrin nor insulin.

The nutritional and hormonal requirements in serum-free medium of normal, phytohemagglutinin-treated lymphocytes isolated from fresh venous blood also have been studied [4, 37, 87]. Human cells in a rich basal nutrient medium consisting of a one-to-one mixture of Dulbecco-modified Eagle's medium and Ham's F12 medium are mitogenically stimulated to grow by the addition of insulin as a hormonal supplement and by ethanolamine, selenium, and transferrin, as nutritional supplements [87]. The cells in this medium grow more slowly than cells in conventional, serum-supplemented medium, representing a reduced number of cycling cells among the viable cell population. Further stimulation of cell growth is observed upon addition of a source of fatty acid in an albumin complex. Estradiol also is reported to cause a small increase in tritiated thymidine incorporation for these cells under serum-free conditions. A number of growth-inhibitory factors have been identified in this assay. These include triiodothyronine, epidermal growth factor and prostaglandin $F_{2\alpha}$.

Conclusions

Recent advances in cell culture techniques allow the maintenance in vitro of many different cell types under conditions in which the extracellular environment is well-defined, compared to the environment created by conventional, serum-supplemented culture media. These technical advances have allowed the design of experiments that not only have identified critical nutritional and hormonal components that could not easily be identified previously, but also have allowed the design of experiments aimed at understanding the mechanisms of action of these factors [8, 9, 11, 12, 14, 54, 81, 86, 116, 117]. Perhaps more striking than the large number of different factors found to be required for the growth in vitro of many cell lines are the large number of different cell types found to respond to a few of these hormonal and nutritional factors, such as insulin, transferrin, lipids, particularly in the form of fatty acids, and trace elements, particularly selenium. The use of precise conditions for the maintenance and growth of cells in culture should in the future allow rapid progress toward answering the many questions surrounding these observations, as well as providing avenues for new observations and insights on the nutritional and hormonal requirements of animal cells in culture.

References

1　Allegra, J.C.; Lippman, M.E.: Growth of a human breast cancer cell line in serum-free hormone-supplemented medium. Cancer Res. *38:* 3823–3829 (1978).

2　Ambesi-Impiombato, F.S.; Parks, L.A.M.; Coon, H.G.: Culture of hormone-dependent functional epithelial cells from rat thyroids. Proc. natn. Acad. Sci. USA *77:* 3455–3459 (1980).

3　Anderson, L.C.; Jokinen, M.; Gahmberg, C.G.: Induction of erythroid differentiation in the human cell line K-562. Nature, Lond. *278:* 364–366 (1979).

4　Arai, S.; Yamane, J.; Tanno, Y.; Takishima, T.: Role of bovine serum albumin in blastoid transformation of lymphocytes by phytohemagglutinin. Proc. Soc. exp. Biol. Med. *154:* 444–448 (1977).

5　Barnes, D.W.: Epidermal growth factor inhibits growth of A431 human epidermoid carcinoma in serum-free cell culture. J. Cell Biol. *93:* 1–4 (1982).

6　Barnes, D.W.: Growth of A-431 human epidermoid carcinoma in serum-free cell culture: inhibition by epidermal growth factor. Cold Spring Harbor Conf. on Cell Proliferation, vol. 9, pp. 937–941 (Cold Spring Harbor Laboratory, New York 1982).

7　Barnes, D.: Growth characteristics of A431 human epidermoid carcinoma cells in serum-free medium: inhibition by epidermal growth factor. Adv. exp. Med. Biol., vol. 172, pp. 49–66 (Plenum Publishing, New York 1984).

8　Barnes, D.: Hormonally defined, serum-free media for epithelial cells in culture; in Taub, Tissue culture in the study of epithelial transport (Plenum Publishing, New York 1984).

9　Barnes, D.W.; Masui, H.: Human cell cultures and serum-free medium. Biomedicine *34:* 67–70 (1981).

10　Barnes, D.W.; Sato, G.H.: Growth of a human breast cancer cell line in serum-free medium. Nature, Lond. *281:* 388–389 (1979).

11   Barnes, D.; Sato, G.: Methods for growth of cultured cells in serum-free medium. Analyt. Biochem. *102:* 255–270 (1980).

12   Barnes, D.W.; Sato, G.H.: Serum-free cell culture: a unifying approach. Cell *22:* 649–655 (1980).

13   Barnes, D.; Sato, G.: Factors that stimulate proliferation of breast cancer cells in vitro in serum-free medium; in McGrath, Brennan, Rich, Cell biology of breast cancer, pp. 277–287 (Academic Press, New York 1980).

14   Barnes, D.; Sato, G.: A general method for cell culture; in Mahesh, Functional correlates of hormone receptors in reproduction, pp. 3–11 (Elsevier, New York 1980).

15   Barnes, D.; Orly, J.; Darmon, M.: Serum-spreading factor: effects on RF1 rat ovary cells and 1003 mouse embryonal carcinoma cells in serum-free media. Cold Spring Harbor Conf. on Cell Proliferation, vol. 9, pp. 155–167 (Cold Spring Harbor Laboratory, New York 1981).

16   Barnes, D.W.; Bosch, J. van der; Masui, H.; Miyzaki, K.; Sato, G.: The culture of human tumor cells; in Pestka, Methods in Enzymology, vol. 79, pp. 368–391 (Academic Press, New York 1981).

17   Barnes, D.; Wolfe, R.; Serrero, G.; McClure, D.; Sato, G.: Effects of serum spreading factor on growth and morphology of cells in serum-free media. J. supramol. Struct. *14:* 47–64 (1980).

18   Bottenstein, J.: Serum-free culture of neuroblastoma cells; in Evans, Advances in neuroblastoma research, pp. 161–170 (Raven Press, New York 1980).

19   Bottenstein, J.: Differentiated properties of neuronal cell lines; in Sato, Functionally differentiated cell lines, pp. 155–184 (Liss, New York 1981).

20   Bottenstein, J.: Growth requirements of neural cells in vitro. Adv. Cell Neurobiol. *4:* 333–379 (1983).

21   Bottenstein, J.; Hayashi, I.; Hutchings, S.; Masui, H.; Mather, J.; McClure, D.B.; Ohasa, S.; Rizzino, A.; Sato, G.; Serrero, G.; Wolfe, R.; Wu, R.: The growth of cells in serum-free hormone-supplemented media. Meth. Enzym. *58:* 94–109 (1979).

22   Bottenstein, J.E.; Sato, G.H.: Growth of a rat neuroblastoma cell line in serum-free medium. Proc. natn. Acad. Sci. USA *76:* 514–517 (1979).

23   Bottenstein, J.E.; Skaper, S.D.; Varon, S.S.; Sato, G.H.: Selective survival of neurons from chick embryo sensory ganglionic dissociates utilizing serum-free supplemented medium. Expl Cell Res. *125:* 183–190 (1980).

24   Breitman, T.R.: Induction of terminal differentiation of HL-60 and fresh leukemic cells by retinoic acid; in Revoltella et al., Expression of differentiated functions in cancer cells, pp. 257–273 (Raven Press, New York 1982).

25   Breitman, T.R.; Collins, S.J.; Keene, B.R.: Replacement of serum by insulin and transferrin supports growth and differentiation of the human promyelocytic cell line, HL-60. Expl Cell Res. *126:* 494–498 (1980).

26   Breitman, T.R.; Collins, S.J.; Keene, B.R.: Terminal differentiation of human promyelocytic leukemic cells in primary culture in response to retinoic acid. Blood *57:* 1000–1004 (1981).

27   Breitman, T.R.; Selonick, S.E.; Collins, S.J.: Induction of differentiation of the human promyelocytic leukemic cell line (HL-60) by retinoic acid. Proc. natn. Acad. Sci. USA *77:* 2936–2940 (1980).

28   Broome, J.D.; Jeng, M.W.: Promotion of replication in lymphoid cells by specific thiols and disulfides, in vitro. J. exp. Med. *138:* 574–592 (1973).

29 Brown, M.; Goldstein, J.: The low density lipoprotein pathway and its relation to atherosclerosis. Am. Rev. Biochem. *46:* 897–930 (1978).

30 Carney, D.N.; Bunn, P.A.; Gazdar, A.F.; Pagan, J.A.; Minna, J.D.: Selective growth in serum-free hormone-supplemented medium of tumor cells obtained by biopsy from patients with small cell carcinoma of the lung. Proc. natn. Acad. Sci. USA *78:* 3185–3189 (1981).

31 Carney, D.N.; Gazdar, A.F.; Minna, J.D.: Positive correlation between histological tumor involvement and generation of tumor cell colonies in agarose in specimens taken directly from patients with small-cell carcinoma of the lung. Cancer Res. *40:* 1820–1823 (1980).

32 Chaproniere-Rickenberg, D.M.; Webber, M.M.: A chemically defined medium for the growth of adult human prostatic epithelium; in Sato, Pardee, Sirbasku, Growth of cells in hormonally defined media, vol. 9, book B, pp. 1109–1115 (Cold Spring Harbor Laboratory, New York 1982).

33 Cherington, P.V.; Smith, B.L.; Pardee, A.B.: Loss of epidermal growth factor requirement and malignant transformation. Proc. natn. Acad. Sci. USA *76:* 3937–3941 (1979).

34 Chuman, L.; Fine, L.G.; Cohen, A.I.; Saier, M.H.: Continuous growth of proximal tubular kidney epithelial cells in hormone-supplemented serum-free medium. J. Cell Biol. *94:* 506–510 (1982).

35 Chung, S.D.; Alaui, N.; Livingston, D.; Hiller, S.; Taub, M.: Characterization of primary rabbit kidney cultures that express proximal tubule functions in hormonally defined medium. J. Cell Biol. *95:* 118–126 (1982).

36 Cioe, L.; McNab, A.; Hubbell, H.R.; Meo, P.; Curtis, P.; Rovera, G.: Differential expression of the globin genes in human leukemia K562 cells induced to differentiate by hemin or butyric acid. Cancer Res. *41:* 237–243 (1981).

37 Darfler, F.J.; Insel, P.A.: Serum-free culture of resting PHA-stimulated, and transformed lymphoid cells inducing hybridomas. Expl Cell Res. *138:* 287–295 (1982).

38 Darfler, F.J.; Insel, P.A.: Clonal growth of lymphoid cells in serum-free media requires elimination of H_2O_2 toxicity. J. cell. Physiol. *115:* 31–36 (1983).

39 Darfler, F.J.; Murakami, H.; Insel, P.A.: Growth of T-lymphoma cells in serum-free medium: lack of involvement of the cyclic AMP pathway in long-term cultures. Proc. natn. Acad. Sci. USA *77:* 5993 (1980).

40 Darmon, M.; Bottenstein, J.; Sato, G.: Neural differentiation following culture of embryonal carcinoma cells in a serum-free defined medium. Devl Biol. *85:* 463–473 (1981).

41 Darmon, M.; Serrero, G.; Rizzino, A.; Sato, G.: Isolation of myoblastic, fibroadipogenic, and fibroblastic clonal cell lines from a common precursor and study of their requirements for growth and differentiation. Expl Cell Res. *132:* 313–327 (1981).

42 Darmon, M.; Stallcup, W.B.; Pittman, Q.J.: Induction of neural differentiation by serum deprivation in cultures of the embryonal carcinoma cell line 1003. Expl Cell Res. *138:* 73–78 (1982).

43 Ellis, A.V.; Demetrius, E.; Temple, T.; Jaffee, R.: Metanephric development in serum-free organ culture. In Vitro *18:* 675–682 (1982).

44 Faivre-Bauman, A.; Rosenbaum, E.; Puymirat, J.; Grouselle, D.; Tixier-Vidal, A.: Differentiation of fetal mouse hypothalamic cells in serum-free medium. Dev. Neurosci. *4:* 118–125 (1981).

45 Ferrero, D.; Tasella, C.; Gallo, E.; Ruscett, F.W.; Breitman, T.R.: Terminal differentiation of the human promyelocytic leukemia cell line, HL-60, in the absence of cell proliferation. Cancer Res. *42:* 4421–4426 (1982).

46 Gallagher, R.; Collins, S.; Trujillo, J.; McCredie, K.; Ahearn, M.; Tsai, S.; Metzgar, R.; Aulakh, G.; Ting, R.; Ruscetti, F.; Gallo, R.: Characterization of the continuous differentiating myeloid cell line (HL-60) from a patient with acute promyelocytic leukemia. Blood *54:* 713–733 (1979).

47 Gill, G.N.; Hornsby, P.J.; Simonian, M.H.: Regulation of growth and differentiated function of bovine adrenocortical cells; in Sato, Ross, Cold Spring Harbor Conf. on Cell Proliferation; Hormones in Cell Culture, vol. 6, pp. 701–715 (Cold Spring Harbor Laboratory, New York 1979).

48 Goodman, R.: Growth and differentiation of pheochromocytoma cells in chemically defined medium; in Barnes, Sirbasku, Sato, Cell culture methods for molecular and cell biology, vol. 4 (Liss, New York 1984).

49 Gospodarowicz, D.: Preparations and uses of lipoproteins to culture normal diploid and tumor cells under serum-free conditions; in Barnes, Sirbasku, Sato, Cell culture methods for molecular and cell biology, vol. 4 (Liss, New York 1984).

50 Gospodarowicz, D.; Lui, G.M.; Gonzalez, R.: High-density lipoproteins and the proliferation of human tumor cells maintained on extracellular matrix-coated dishes and exposed to defined medium. Cancer Res. *42:* 3704–3713 (1982).

51 Guilbert, L.J.; Iscove, N.N.: Partial replacement of serum by selenite, transferrin, albumin and lecithin in haemopoietic cell cultures. Nature, Lond. *263:* 594–595 (1976).

52 Ham, R.G.: An improved nutrient solution for diploid Chinese hamster and human cell lines. Expl Cell Res. *29:* 515–526 (1963).

53 Ham, R.G.: Survival and growth requirements of nontransformed cells; in Baserga, Handbook of experimental pharmacology, pp. 13–88 (Springer, New York 1981).

54 Ham, R.G.: Importance of the basal nutrient medium in the design of hormonally defined media; in Sato, Pardee, Sirbasku, Cold Spring Harbor Conf. on Cell Proliferation, vol. 9, pp. 39–60 (Cold Spring Harbor Laboratory, New York 1982).

55 Ham, R.G.; McKeehan, W.L.: Nutritional requirements for clonal growth of nontransformed cells; in Katsuta, Nutritional requirements of cultured cells, pp. 63–115 (Jap. Sci. Soc. Press, Tokyo 1978).

56 Ham, R.G.; McKeehan, W.L.: Media and growth requirements. Meth. Enzym. *58:* 44–95 (1979).

57 Havel, R.J.; Eder, H.A.; Bragdon, J.H.: The distribution and chemical composition of ultracentrifugally separated lipoproteins in human serum. J. clin. Invest. *34:* 1345–1353 (1955).

58 Hayashi, I.; Sato, G.H.: Replacement of serum by hormones permits growth of cells in a defined medium. Nature, Lond. *259:* 132–134 (1976).

59 Hayashi, I.; Sato, G.H.; Larner, J.: Hormonal growth control of cells in culture. In Vitro *14:* 23–30 (1978).

60 Higuchi, K.: Cultivation of animal cells in chemically defined media, a review. Adv. appl. Microbiol. *16:* 111–160 (1973).

61 Honma, Y.; Kusakabe, T.; Okabe, J.; Hozumi, M.: Replacement of serum by insulin, transferrin, albumin, phosphatidylcholine, cholesterol and some trace elements in cultures of mouse myeloid leukemia cells sensitive to inducers of differentiation. Expl Cell Res. *124:* 421 (1979).

62   Hutchings, S.E.; Sato, G.H.: Growth and maintenance of HeLa cells in serum-free medium supplemented with hormones. Proc. natn. Acad. Sci. USA *75:* 901–904 (1978).

63   Imagawa, W.; Tomooka, Y.; Nandi, S.: Serum-free growth of normal and tumor mouse mammary epithelial cells in primary culture. Proc. natn. Acad. Sci. USA *79:* 4074–4077 (1982).

64   Iscove, N.N.; Guilbert, L.J.; Weyman, C.: Complete replacement of serum in primary cultures of erythroprotein-dependent red cell precursors (CFV-E) by albumin, transferrin, iron, unsaturated fatty acid, lecithin and cholesterol. Expl Cell Res. *126:* 121–126 (1980).

65   Iscove, N.N.; Melchers, F.: Complete replacement of serum by albumin, transferrin, and soybean lipid in cultures of lipopolysaccharide-reactive B lymphocytes. J. exp. Med. *147:* 923–933 (1978).

66   Jetten, A.M.; Jetten, M.E.R.; Sherman, M.I.: Stimulation of differentiation of several murine embryonal carcinoma cell lines by retinoic acid. Expl Cell Res. *124:* 381–391 (1979).

67   Kano-Sueoka, T.; Cohen, D.M.; Yamaizumi, Z.; Nishimura, S.; Mori, M.; Fujiki, H.: Phosphoethanolamine as a growth factor of a mammary carcinoma cell line of rat. Proc. natn. Acad. Sci. USA *76:* 5741–5744 (1979).

68   Kano-Sueoka, T.; Errick, J.E.: Effects of phosphoethanolamine and ethanolamine on growth of mammary carcinoma cells in culture. J. exp. Cell Res. *136:* 137–145 (1981).

69   Kano-Sueoka, T.; Hsieh, P.: A rat mammary carcinoma in vivo and in vitro: establishment of clonal lines of the tumor. Proc. natn. Acad. Sci. USA *70:* 1922–1926 (1973).

70   Koch, K.S.; Leffert, H.L.: Growth control of differentiated adult rat hepatocytes in primary culture. Ann. N.Y. Acad. Sci. *349:* 111–127 (1980).

71   Lechner, J.F.; Haugen, A.; McClendon, I.A.; Pettis, E.W.: Clonal growth of normal adult human bronchial epithelial cells in a serum-free medium. In Vitro *18:* 633–642 (1982).

72   Leffert, H.L.: Growth control of differentiated fetal rat hepatocytes in primary monolayer culture. VII. Hormonal control of DNA synthesis and its possible significance to the problem of liver regeneration. J. Cell Biol. *62:* 792–801 (1974).

73   Loudes, C.; Faivre-Bauman, A.; Barret, A.; Grouselle, D.; Puymirat, A.; Tixier-Vidal, A.: Functional maturation of TRH immunoreactive neurons in serum-free cultures of mouse hypothalamic cells. Dev. Brain Res. (1984).

74   Mather, J.P.: The establishment and characterization of two distinct mouse testicular epithelial cell lines. Biol. Reprod. *23:* 243–250 (1980).

75   Mather, J.P.; Haour, F.: Hormone response of testicular cells in culture: established cell lines and primary cultures; in Sato, Functionally differentiated cell lines, pp. 93–108 (Liss, New York 1981).

76   Mather, J.P.; Saez, J.M.; Dray, F.; Haour, F.: Hormone-hormone and hormone-vitamin interactions in the control of growth function of Leydig cells in vitro. Cold Spring Harbor Conf. on Cell Proliferation, vol. 9, pp. 1117–1128 (Cold Spring Harbor Laboratory, New York 1982).

77   Mather, J.P.; Saez, J.M.; Dray, F.; Haour, F.: Vitamin E prolongs survival and function of porcine Leydig cells in culture. Acta endocr., Copenh. *102* (1983).

78 Mather, J.P.; Saez, J.M.; Haour, F.: Primary cultures of Leydig cells for rat, mouse and pig: advantages of porcine cells for the study of gonadotropin regulation of Leydig cell function. Steroids *38:* 35–44 (1981).

79 Mather, J.P.; Saez, J.M.; Haour, F.: Regulation of gonadotropin receptors and steroidogenesis in cultured porcine Leydig cells. Endocrinology *110:* 933–940 (1982).

80 Mather, J.P.; Sato, G.H.: The growth of mouse melanoma cells in hormone-supplemented, serum-free medium. Expl Cell Res. *120:* 191–197 (1979).

81 Mather, J.P.; Sato, G.H.: The use of hormone-supplemented serum-free media in primary cultures. Expl Cell Res. *124:* 215–221 (1979).

82 Mather, J.P.; Zhuang, L.Z.; Perez-Infante, V.; Phillips, D.M.: Culture of testicular cells in hormone-supplemented serum-free medium. Ann. N.Y. Acad. Sci. *383:* 44–68 (1982).

83 McClure, D.B.: Anchorage-independent colony formation of SV40 transformed BALBc-3T3 cells in serum-free medium: role of cell- and serum-derived factors. Cell *32:* 999–1006 (1983).

84 McClure, D.B.; Hightower, M.J.; Topp, W.C.: Effect of SV40 transformation on the growth-factor requirements of the rat embryo cell line REF52 in serum-free medium; in Sirbasku, Sato, Pardee, Growth of cells in hormonally defined medium, pp. 345–364 (Cold Spring Harbor Laboratory, New York 1982).

85 McKeehan, W.L.: Growth-factor-nutrient interrelationships in control of normal and transformed cell proliferation; in Sato, Pardee, Sirbasku, Cold Spring Harbor Conf. on Cell Proliferation, vol. 9, pp. 65–74 (Cold Spring Harbor Laboratory, New York 1982).

86 McKeehan, W.; Hamilton, W.G.; Ham, R.G.: Selenium is an essential trace nutrient for the growth of WI-38 diploid human fibroblasts. Proc. natn. Acad. Sci. USA *73:* 2023–2027 (1976).

87 Mendelsohn, J.; Caviles, A., Jr.; Castagnola, J.: Proliferation of normal human lymphocytes in serum-free albumin-free medium; in Sirbasku, Sato, Pardee, Growth of cells in hormonally defined medium, pp. 677–690 (Cold Spring Harbor Laboratory, New York 1982).

88 Minna, J.D.; Carney, D.N.; Oie, H.; Bunn, P.A.; Gazdar, A.F.: Growth of human small-cell lung cancer in defined medium; in Pardee, Sato, Growth of cells in hormonally defined medium, pp. 627–639 (Cold Spring Harbor Laboratory, New York 1982).

89 Miyazaki, K.; Masui, H.; Sato, G.: Control factors for keratinization of human bronchogenic epidermoid carcinoma cells; in Sato, Pardee, Sirbasku, Growth of cells in hormonally defined media, vol. 9, book B, pp. 657–661 (Cold Spring Harbor Laboratory, New York 1982).

90 Morrison, R.S.; deVellis, J.: Growth of purified astrocytes in chemically defined medium. Proc. natn. Acad. Sci. USA *78:* 7205–7209 (1980).

91 Morrison, R.S.; deVellis, J.: Growth and differentiation of purified astrocytes in a chemically defined medium; in Sirbasku, Sato, Pardee, Growth of cells in hormonally defined medium, pp. 627–639 (Cold Spring Harbor Laboratory, New York 1982).

92 Murakami, H.; Masui, H.: Hormonal control of human colon carcinoma cell growth in serum-free medium. Proc. natn. Acad. Sci. USA *77:* 3464–3468 (1980).

93 Murakami, H.; Masui, H.; Sato, G.; Roschke, W.C.: Growth of mouse plasmacytoma cells in serum-free hormone-supplemented medium: procedure for the determination of hormone and growth factor requirements for cell growth. Analyt. Biochem. *114:* 422 (1981).

94 Murakami, H.; Masui, H.; Sato, G.H.; Sueoka, N.; Chow, T.P.; Kano-Sueoka, T.: Growth of hybridoma cells in serum-free medium: ethanolamine is an essential component. Proc. natn. Acad. Sci. USA *79:* 1158–1162 (1982).

95 Nilausen, K.: Role of fatty acids in growth-promoting effect of serum albumin on hamster cells in vitro. J. cell. Physiol. *96:* 1–14 (1978).

96 Oh, S.H.; Ganther, H.E.; Hoekstra, W.G.: Selenium as a component of glutathione peroxidase isolated from ovine erythrocytes. Biochemistry, N.Y. *13:* 1825–1829 (1974).

97 Orly, J.; Sato, G.: Fibronectin mediates cytokinesis and growth of rat follicular cells in serum-free medium. Cell *17:* 295–305 (1979).

98 Orly, J.; Sato, G.; Erickson, G.F.: Serum suppresses the expression of hormonally induced functions in cultured granulosa cells. Cell *20:* 817–827 (1980).

99 Peehl, D.M.; Ham, R.G.: Clonal growth of human keratinocytes with small amounts of dialyzed serum. In Vitro *16:* 526–538 (1980).

100 Perez-Infante, V.; Mather, J.P.: The role of transferrin in the growth of testicular cell lines in serum-free medium. Expl Cell Res. *142:* 325–332 (1982).

101 Phillips, J.L.; Azari, P.: Zinc transferrin. Enhancement of nucleic acid synthesis in phytohemagglutinin-stimulated human lymphocytes. Cell. Immunol. *10:* 31–37 (1974).

102 Pickart, L.; Freedman, J.H.; Loker, W.J.; Peisach, J.; Perkins, C.M.; Stenkamp, R.E.; Weinstein, B.: Growth-modulating plasma tripeptide may function by facilitating copper uptake into cells. Nature, Lond. *228:* 715–717 (1980).

103 Pipas, J.M.; Chiang, L.C.; Barnes, D.W.: Influence of genetic factors on simian virus 40-induced neoplastic transformation. Cold Spring Harbor Conf. on Cell Proliferation, vol. 11 (Cold Spring Harbor Laboratory, New York 1983).

104 Pledger, W.J.; Leof, E.B.; Chou, N.; Olashaw, E.J.; O'Keefe, E.J.; Wyk, J.J. von; Wharton, W.R.: Initiation of cell-cycle traverse by serum-derived growth factors; in Sirbasku, Sato, Pardee, Growth of cells in hormonally defined medium, pp. 259–273 (Cold Spring Harbor Laboratory, New York 1982).

105 Puymirat, J.; Barret, A.; Picart, R.; Vigny, A.; Loudes, C.; Faivre-Bauman, A.; Tixier-Vidal, A.: Triiodothyronine enhances the morphological maturation of dopaminergic neurons in serum-free medium cultures of fetal mouse hypothalamic cells. Neuroscience *10:* 801–810 (1983).

106 Reid, L.; Sato, G.: Replacement of serum in cell culture by hormones; in Rickenberg, Biochemistry and mode of action of hormones, vol. 11, pp. 219–251 (University Park Press, Baltimore 1978).

107 Reznikoff, C.A.; Brankow, D.W.; Heidelberger, C.: Establishment and characterization of a cloned line of C3H mouse embryo cells sensitive to postconfluence inhibition of division. Cancer Res. *33:* 3231–3238 (1973).

108 Rizzino, A.: Two multipotent embryonal carcinoma cell lines irreversibly differentiated in defined media. Devl Biol. 95: 126–136 (1983).

109 Rizzino, A.; Crowley, C.: Growth and differentiation of embryonal carcinoma cell line F9 in defined media. Proc. natn. Acad. Sci. USA *77:* 457–461 (1980).

110 Rizzino, A.; Rizzino, H.; Sato, G.: Defined media and the determination of nutritional and hormonal requirements of mammalian cells in culture. Nutr. Rev. *37:* 369–378 (1979).

111 Rizzino, A.; Sato, G.: Growth of embryonal carcinoma cells in serum-free medium. Proc. natn. Acad. Sci. USA *75:* 1844–1848 (1978).

112 Rizzino, A.; Sherman, M.I.: Blastocysts in serum-free medium. Expl Cell Res. *121:* 221–233 (1979).

113 Rizzino, A.; Terranova, V.; Rohrbach, D.; Crowley, C.; Rizzino, H.: The effects of laminin on the growth and differentiation of embryonal carcinoma cells in defined media. J. supramol. Struct. *13:* 243–253 (1980).

114 Rockwell, G.A.; McClure, D.; Sato, G.H.: The growth requirements of SV40 virus-transformed Balb/C-3T3 cells in serum-free monolayer culture. J. cell. Physiol. *103:* 323–331 (1980).

115 Saier, M.H., Jr.: Hormonally defined, serum-free medium for proximal tubular kidney epithelial cell line, LLC-PK1; in Barnes, Sirbasku, Sato, Methods for serum-free cell culture (Liss, New York 1983).

116 Sato, G.H.: The role of serum in cell culture; in Litwack, Biochemical actions of hormones, pp. 391–396 (Academic Press, New York 1975).

117 Sato, G.; Pardee, A.B.; Sirbasku, D.A. (eds): Cold Spring Harbor Conf. on Cell Proliferation, vol. 9 (Cold Spring Harbor Laboratory, New York 1982).

118 Simms, E.; Gazdar, A.F.; Abrams, P.G.; Minna, J.D.: Growth of human small cell (oat cell) carcinoma of the lung in serum-free growth of factor-supplemented medium. Cancer Res. *40:* 4356–4363 (1980).

119 Simonian, M.H.; White, M.L.; Gill, G.N.: Growth and function of cultured bovine adrenocortical cells in a serum-free defined medium. Endocrinology *111:* 919–927 (1982).

120 Solomon, D.S.; Liotta, L.A.; Kidwell, W.R.: Differential response to growth factor by rat mammary epithelium plated on different collagen substrata in serum-free medium. Proc. natn. Acad. Sci. USA *78:* 382–386 (1981).

121 Spector, A.: Fatty acid binding to plasma albumin. J. Lipid Res. *16:* 165–179 (1975).

122 Spieker-Polet, H.; Polet, H.: Requirement of a combination of a saturated and an unsaturated free fatty acid and a fatty acid carrier protein for in vivo growth of lymphocytes. J. Immun. *126:* 946–964 (1981).

123 Stampfer, M.: Methods for growth of human mammary epithelial cells in monolayer culture; in Barnes, Sirbasku, Sato, Methods for serum-free cell culture (Liss, New York 1983).

124 Stampfer, A.; Hallowes, R.C.; Hackett, A.D.: Growth of normal human mammary cells in culture. In Vitro *16:* 415–423 (1980).

125 Strickland, S.; Mahdavi, V.: The induction and differentiation in teratocarcinoma stem cells by retinoic acid. Cell *15:* 393–403 (1978).

126 Strickland, S.; Smith, K.K.; Marotti, K.R.: Hormonal induction of differentiation in teratocarcinoma stem cells: generation of parietal endoderm by retinoic acid and dibutyryl cAMP. Cell *21:* 347–355 (1980).

127 Takemoto, H.; Yokoro, K.; Furth, J.; Cohen, A.J.: Adrenotropic activity of mammosomatotropic tumors in rats and mice. Cancer Res. *22:* 917–924 (1962).

128 Taub, M.; Chuman, L.; Saier, M.H.; Sato, G.: Growth of Madin Darby canine kidney

epithelial cell (MDCK) line in hormone-supplemented serum-free medium. Proc. natn. Acad. Sci. USA *76:* 3338–3342 (1979).

129 Taub, M.; Livingston, D.: The development of serum-free hormone-supplemented media for primary kidney cultures and their use in examining renal functions. Ann. N.Y. Acad. Sci. *372:* 406–421 (1981).

130 Taub, M.; Saier, M.H., Jr.; Chuman, L.; Hiller, S.: Loss of the PGE_1 requirement for MDCK cell growth associated with a defect in cyclic AMP phosphodiesterase. J. cell. Physiol. *114:* 153–161 (1983).

131 Taub, M.; Sato, G.: Growth of kidney epithelial cells in hormone-supplemented, serum-free medium. J. supramol. Struct. *11:* 207–216 (1979).

132 Taub, M.; Sato, G.: Growth of functional primary cultures of kidney epithelial cells in defined medium. J. cell. Physiol. *105:* 369–378 (1980).

133 Taub, M.; Wu, B.; Chuman, L.; Rindler, M.J.; Saier, M.H., Jr.; Sato, G.: Alterations in growth requirements of kidney epithelial cells in defined medium associated with malignant transformation. J. supramol. Struct. *15:* 63–72 (1981).

134 Tauber, J.P.; Cheng, J.; Massoglia, S.; Gospodarowicz, D.: High density lipoproteins and the growth of vascular endothelial cells in serum-free medium. In Vitro *17:* 519–530 (1981).

135 Tormey, D.C.; Mueller, G.C.: Biological effects of transferrin on human lymphocytes in vitro. Expl Cell Res. *74:* 220 (1972).

136 Tsao, M.C.; Walthall, B.J.; Ham, R.G.: Clonal growth of normal human epidermal keratinocytes in a defined medium. J. cell. Physiol. *110:* 219 (1982).

137 Van der Bosch, J.: Primary tissue cultures of human colon carcinomas in serum-free medium: an in vitro system for tumor analysis and therapy experiments; in Barnes, Sirbasku, Sato, Methods for serum-free cell culture (Liss, New York 1983).

138 Van der Bosch, J.; Masui, H.; Sato, G.: Growth characteristics of primary tissue cultures from heterotransplanted human colorectal carcinomas in serum-free medium. Cancer Res. *41:* 611–618 (1981).

139 Varga, J.M.; Lambert, D.T.; Airoldi, L.; Moellmann, G.E.; Hudson, A.; Yu, R.K.; Makala, L.J.; Kuklinska, E.; Minsner, P.; Lerner, A.B.; Bartholomew, J.C.: Phenotypic variations in murine melanoma cells grown in serum-containing and defined media; in Sirbasku, Sato, Pardee, Growth of cells in hormonally defined medium, pp. 899–910 (Cold Spring Harbor Laboratory, New York 1982).

140 Walthall, B.J.; Ham, R.G.: Multiplication of human diploid fibroblasts in a synthetic medium supplemented with EGF, insulin and dexamethasone. Expl Cell Res. *134:* 301 (1981).

141 Wang, R.J.; Nixon, B.T.: Identification of hydrogen peroxide as a photoproduct toxic to human cells in tissue-culture medium irradiated with 'daylight' fluorescent light. In Vitro *14:* 715–722 (1978).

142 Weinstein, R.; Hoover, G.A.; Majure, J.; van der Spek, J.; Stemerman, M.B.; Maciag, T.: Growth of human foreskin fibroblasts in a serum-free, defined medium without platelet-derived growth factor. J. cell. Physiol. *110:* 23–28 (1982).

143 Wolfe, R.A.; Sato, G.H.; McClure, D.B.: Continuous culture of rat C6 glioma in serum-free medium. J. Cell Biol. *87:* 434–441 (1980).

144 Wolfe, R.A.; Wu, R.; Sato, G.: EGF-induced down regulation of receptor does not occur in HeLa cells grown in defined medium. Proc. natn. Acad. Sci. USA *77:* 2735–2739 (1980).

145 Wu, R.; Sato, G.H.: Replacement of serum in cell culture by hormones: a study of hormonal regulation of cell growth and specific gene expression. J. Toxicol. envir. Hlth *4:* 427–448 (1978).
146 Yamane, I.; Kan, M.; Minamoto, Y.; Amatsuji, Y.: α-Cyclodextrin, a novel substitute for bovine albumin in serum-free culture of mammalian cells. Proc. Jap. Acad. *57:* 385 (1981).
147 Yamane, I.; Murakami, O.; Kato, M.: 'Serum-free' culture of various mammalian cells and the role of bovine albumin. Cell Struct. Funct. *1:* 279–284 (1976).
148 Yang, J.; Larson, L.; Flynn, D.; Elias, J.; Nandi, S.: Serum-free primary culture of human normal mammary epithelial cells in collagen gel matrix. Cell Biol. int. Rep. *6:* 969–975 (1982).
149 Yasumura, Y.; Tashjian, A.H., Jr.; Sato, G.: Establishment of four functional clonal strains of animal cells in culture. Science *154:* 1186–1189 (1966).

D. Barnes, PhD, Department of Biological Sciences, University of Pittsburgh, Pittsburgh, PA 15260 (USA)

Wld Rev. Nutr. Diet., vol. 45, pp. 198–220 (Karger, Basel 1985)

Mechanism of Conversion of
β-Carotene into Vitamin A –
Central Cleavage versus Random Cleavage

J. Ganguly[1], P.S. Sastry

Department of Biochemistry, Indian Institute of Science, Bangalore, India

Contents

I. Introduction

After the existence of fat-soluble vitamin A was well recognized and accepted, considerable confusion arose during 1920s regarding two quite different types of substances showing vitamin A activity in animals. To one group belonged the highly coloured hydrocarbon compounds extensively found in the plant kingdom and collectively called carotenoids, while the

---

[1] Prof. *J. Ganguly* is an Emeritus Medical Scientist of the Indian Council of Medical Research, New Delhi, India.

other compound is almost colourless and is concentrated in the liver. This confusion was ultimately resolved from the biological work of *Moore* [1930] and the chemical work of *Karrer* et al. [1930, 1931]. *Moore* [1930] demonstrated that rats fed massive amounts of carotenoids extracted from carrots deposited large amounts of vitamin A in their livers. At about the same time the classical work of *Karrer* et al. [1930, 1931] decisively established the chemical relationship between β-carotene and vitamin A, and it became clear that the carotene acts as a precursor for vitamin A and is converted to vitamin A in the animal body. The structures of the two compounds led *Karrer* to suggest that simple addition of two molecules of water at the central double bond of the carotene molecule should give rise to two molecules of vitamin A. But eventually it turned out that the biological conversion of carotene into vitamin A is not so simple. Actually, two theories have been put forward regarding the mechanism of conversion. According to one of them, the carotene is cleaved asymmetrically (random cleavage theory), while the other has suggested that the attack is at the central double bond of the carotene molecule (central cleavage theory). The purpose of this review is to point out that, although 50 years have elapsed since the relationship between carotene and vitamin A was established, our knowledge regarding the actual mechanism of biological transformation of carotene into vitamin A is far from clear and that the weight of evidence is overwhelmingly in favour of random cleavage.

II. Attempts at Conversion by the Liver, in vitro

After *Moore* [1930] had demonstrated that feeding of carotene leads to deposition of vitamin A in the liver, it was widely believed that the conversion takes place in the liver. In fact, some laboratories [*Olcott and McCann*, 1931] even claimed to have isolated an enzyme from the liver of vitamin A-deficient rats, which on incubation with a carotene preparation yielded a compound with absorption at 325 nm, and the enzyme was called 'carotenase'. But such claims were immediately challenged by *Woolf and Moore* [1932], who pointed out that the mere appearance of a band at 325 nm did not conclusively establish that vitamin A was actually formed and that the product must be convincingly characterized. Eventually reports from other laboratories showed that such criticisms were indeed valid, because similar attempts in these laboratories failed to establish that vitamin A was actually formed in such experiments [*Rea and Drummond*, 1932]. Therefore, at this stage it was generally agreed that enzymic conversion of carotene to vitamin A cannot be demonstrated with liver extracts [see *Ganguly and Murthy*, 1967, for further details].

III. Conversion in the Intestine, in vivo

In spite of such failure to demonstrate the conversion by liver extracts, the belief that the conversion takes place in the liver continued to persist. Quite unexpectedly the whole question took a rather different turn when *Sexton* et al. [1946] reported that vitamin A-deficient rats given intraperitoneal injections of colloidal β-carotene would die of the deficiency, even though large amounts of the injected carotene were found in their livers. On the other hand, when small amounts of carotene were given to similar deficient rats orally, they continued to live. Therefore, these workers came to the only logical conclusion that carotene is converted into vitamin A in the intestine and not in the liver. Soon three groups of workers [*Mattson* et al., 1947; *Glover* et al., 1948a; *Thompson* et al., 1949, 1950] independently demonstrated that, when vitamin A-deficient rats are given orally β-carotene dissolved in oil, vitamin A appears in the intestine long before it appears in the liver. In fact, *Thompson* et al. [1949, 1950] showed that considerable amounts of vitamin A are found in the small intestine within 15 min after the carotene is given to the rats, while the vitamin can be found in the liver much later, generally after about an hour.

These experiments therefore conclusively proved that, when given orally, carotene is converted into vitamin A in the intestine. But soon reports began to appear in the literature claiming that the conversion can take place in other tissues also. Thus, *Bieri and Pollard* [1954] reported that, following intravenous injections of small amounts of β-carotene dispersed in Tween, into vitamin A-deficient rats, vitamin A rapidly appears in the liver and kidneys of these rats. The same workers also reported that there was no significant difference in the amounts of vitamin A found in the blood of rats given intravenous injections of aqueous dispersions of β-carotene, after bilateral nephrectomy, partial hepatectomy, ligation of the bile duct or enterectomy. Soon similar findings were reported from other laboratories also, where the liver or liver plus viscera (consisting of the stomach, large and small intestines, pancreas, adrenals, kidneys and gonads) were removed from the rats [*McGillivray* et al., 1956]. All these observations therefore showed that the reaction is not confined to the intestine alone and that it can take place in many other tissues of animals, provided the carotene is made available to the tissues in the form of a suitable aqueous dispersion, so that the tissue enzymes can have easy access to it [see *Ganguly and Murthy*, 1967, for further details].

IV. Conversion, in vitro

A. With Isolated Tissues

It has already been discussed that there were earlier claims of demonstration of conversion, in vitro, by liver extracts, and that such claims were eventually found to be not convincing. But after it was demonstrated that the intestine is the major site of conversion of the orally administered carotene, the question of in vitro conversion was re-opened, and renewed attempts were made in some laboratories to demonstrate such a reaction using whole tissue, or segments, slices or homogenates of tissues. Thus, *Wiese* et al. [1947], who had incubated the small intestine of rats, which were previously given orally β-carotene dissolved in oil, were led to believe that the reaction took place in the isolated tissue. With an almost similar approach *Glover* et al. [1948a] also had found an increase in the vitamin A values of rat intestine. In sharp contrast, *Ganguly* [1949] consistently failed to demonstrate such conversion in rat intestine. Eventually similar attempts in other laboratories confirmed that in vitro conversion cannot be demonstrated in rat liver or intestine [see *Ganguly and Murthy*, 1967, for detailed and critical review of earlier work].

At this point of time, in spite of such claims and counterclaims, it was widely accepted that conversion of carotene into vitamin A cannot be demonstrated in isolated tissues and the situation was reminiscent of the earlier attempts at conversion with liver extracts. But, soon fresh attempts, where radioactive β-carotene began to be used for such work, revealed that the techniques used in the earlier work were not sensitive enough to detect the rather small amounts of vitamin A formed and that by using radiolabelled carotene it is indeed possible to demonstrate in vitro conversion not only in the intestine, but in the liver as well. Thus, *Olson* [1961a] showed that, when aqueous dispersions of ^{14}C-β-carotene in Tween 20 are injected into the intestinal loops of living rats, definite amounts of labelled retinyl esters are found in the intestinal wall, while at the same time small amounts of radioactive retinal, retinol and acidic compounds are also found in the same tissue. Later on *Zachman and Olson* [1963] reported that, following perfusion in isolated rat liver of ^{14}C-β-carotene dispersed in Tween 80 and Krebs-Ringer phosphate buffer, ^{14}C-retinyl esters are formed in the perfused liver. These workers had conclusively identified the product formed after hydrolysis of the ester to retinol, followed by oxidation of the retinol to retinal and formation of the semicarbazone derivative of the aldehyde. During all these steps the specific activities of the derivatives of retinol remained relatively constant.

B. With Enzymes Isolated from Tissues

The interesting experiments with isolated tissues mentioned above were soon followed by work with isolated enzyme systems. Thus, independently and almost simultaneously *Goodman and Olson* claimed demonstration of in vitro conversion of ^{14}C-β-carotene into ^{14}C-retinal by enzymes isolated from the homogenates of liver and intestinal mucosa of rats. *Goodman and Huang* [1965] reported that ^{14}C-retinal is formed when ^{14}C-β-carotene dissolved in traces of acetone is incubated with the 104,000 *g* supernatant and washed cell particles of the homogenates of rat intestinal mucosa. In this particular system molecular oxygen and bile salts were required and the ^{14}C-retinal produced in yields of 30–50% was identified by forming its semicarbazone derivative. Similarly *Olson and Hayaishi* [1965] demonstrated that incubation of the 104,000 *g* supernatant of rat liver homogenates with a micellar solution of ^{14}C-β-carotene leads to the formation of radioactive retinal and retinol. The primary product in this system appeared to be retinal, which was identified by reducing it to retinol by treatment with NaBH$_4$. Here also, the reaction required molecular oxygen, but in contrast to the findings of *Goodman and Huang* [1965] there was no requirement of cell particles or bile salts. In fact, addition of glycocholate to the enzymes prepared from the intestine and kidneys had no effect on the cleavage of β-carotene. These authors could not detect any apo-β-carotenals or carotenols as reaction products and therefore called the enzyme 'β-carotene 15-15′-oxygenase'. They also proposed the mechanism of cleavage of β-carotene at the central double bond, as depicted in figure 1.

Goodman et al. [1966] then showed that during the process of conversion the hydrogen atoms attached to the central carbon atoms of the carotene molecule are not lost. In this work a sample of β-carotene uniformly labelled with ^{14}C throughout the molecule was mixed with another sample of β-carotene which contained ^{3}H only at the central double bond and the mixture was fed to rats with lymph fistula. The ratio of ^{3}H to ^{14}C in the lymph retinyl esters was found to be identical with that of the fed β-caro-

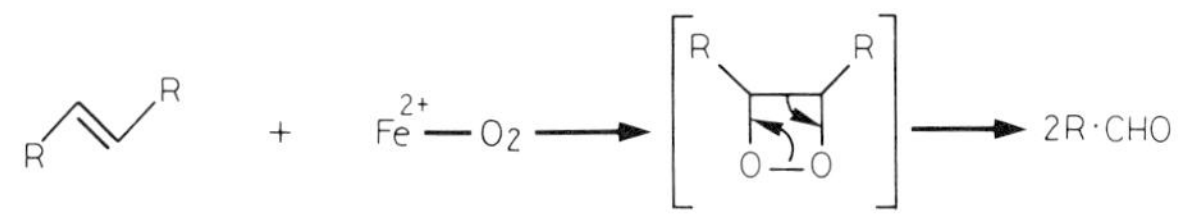

Fig. 1. Proposed mechanism for the cleavage of β-carotene at the central double bond [*Olson and Hayaishi,* 1965]. R denotes the C-19 substituent adjacent to the central double bond.

tene, which indicated that there was no loss of the ^{3}H relative to the ^{14}C during the conversion process. But, when the lymph retinol was oxidized to retinoic acid by a two-step procedure via retinal, complete loss of the ^{3}H from the retinol molecule was observed. It should, however, be clear that these experiments did not establish that the cleavage of the carotene molecule took place at the central double bond only, because if apo-β-carotenals were formed as intermediates they would also eventually lead to the formation of retinol with the same radioactivity.

In another communication *Goodman* et al. [1967] described in detail partial purification of the cleavage enzyme of the 104,000 *g* supernatant fraction of rat mucosal homogenates by 20–45% saturation with ammonium sulphate. Here also, ^{14}C-β-carotene was added in 50 µl of acetone and either cell particles or an appropriate combination of a detergent plus lipid (especially lecithin) was necessary for the reaction, without which the activity of the cleavage enzyme was rather poor.

In yet another report *Fidge* et al. [1969] described a 27-fold purification of the enzyme from hog mucosa successively by 0–55% ammonium sulphate precipitation, DEAE-Sephadex chromatography and gel filtration on Sephadex G-200. Here, the substrate ^{14}C-β-carotene was dispersed in Tween 40 and, according to the investigators, maximal activity could be obtained by the addition of an appropriate combination of detergents and lipids, especially Tween 40, sodium glycocholate and sphingomyelin.

At about the same time *Goodman and Olson* [1969] gave a detailed description of the procedure for preparation and assay of the cleavage enzyme and extensively discussed the requirement of an appropriate detergent-lipid combination like sodium glycocholate plus egg lecithin for the reaction. Apparently, in this system the activity is completely lost without glycocholate.

On the other hand, *Lakshmanan* et al. [1968] precipitated the cleavage enzyme from the 43,500 *g* supernatant fraction of rabbit mucosa by 25–45% saturation with ammonium sulphate and used β-carotene and several apo-β-carotenals as substrates. These substrates were dispersed in sodium lauryl sulphate and no other detergent or lipid, such as lecithin, was used. In this work also retinal was identified as the only product and the rabbit enzyme was markedly more active against 10′-, 8′- and 4′-apo-β-carotenal, as compared to carotene. Essentially similar results were reported by *Olson* [1969].

Soon after these reports appeared *Singh and Cama* [1974] extensively purified the enzymes from the intestines of guinea-pig and rabbits, and used various substrates which included β-carotene and several apocarotenals and apocarotenols. These substrates were dispersed in Tween 20 and no other detergent or lipid like lecithin was used. But in contrast to the findings of *Lakshmanan* et al. [1968] the enzymes from both sources were markedly more active against β-carotene, as compared to the apocarotenals (table I).

The other important observations of *Goodman* [1969], *Olson* [1969] and of *Singh and Cama* [1974] regarding the enzymes from the three species of animals can be summarized as follows: (1) The activity is stimulated by thiol reagents and strongly inhibited by sulphydryl inhibitors as well as by chelating agents like α,α′-dipyridyl and *O*-phenanthrolene. (2) It requires molecular oxygen and it is a dioxygenase reaction, where molecular oxygen reacts with the central two carbon atoms of the β-carotene molecule, after which the central double bond of the carotene is cleaved giving two molecules of retinal.

Requirement of molecular oxygen for the conversion of β-carotene into retinyl esters by isolated intestinal sections was also demonstrated by *Olson* [1961b] and by *Harashima* [1964], while *Vartapetian* et al. [1966] had shown that ^{18}O was incorporated into the liver retinyl esters from molecular oxygen, but not from water, when vitamin A-deficient rats were fed β-carotene.

C. Stoichiometry of the Reaction

Since two alternative mechanisms have been proposed for the cleavage of β-carotene, namely central cleavage and random cleavage, a critical examination of the actual stoichiometry of the reaction appears essential. In their first communication *Goodman and Huang* [1965] had used '^{14}C-β-carotene, produced biosynthetically from ^{14}C-acetate by the fungus *Phycomyces blakesleeanus*', but did not describe the radiopurity of the substrate used. Nevertheless, we assume that the radiopurity of the ^{14}C-β-carotene was satisfactory. Their reaction mixture contained, among others, '11 mg of the soluble protein (104,000 g supernatant), an equivalent amount of washed cell particles (sedimented between 2,000 and 104,000 g)' and ^{14}C-β-carotene (0.7 μg, about 1,100 cpm) in 50 μl of acetone. Regarding the stoichiometry of the reaction, they stated as follows:

'Calculations based on the stoichiometry of our results indicate that most of the reaction-product retinal must have arisen by central cleavage of the substrate β-carotene into two molecules of retinal. Thus, in one experiment three incubations were conducted without enzyme or with inactivated enzyme, and four incubations were conducted simultaneously with active enzyme. The same amount of substrate was added to each of the seven incubation flasks. After incubation, 1,029 ± 13 cpm were recovered in column fraction 1 (carotene) and 55 ± 9 cpm in fraction 3, from the three flasks without active enzyme. After incubation with active enzyme, 535 ± 28 cpm were recovered in fraction 1 and 460 ± 20 cpm in fraction 3. These results are not consistent with the possibility that only one molecule of retinal is formed from one molecule of β-carotene.'

In all subsequent publications *Goodman* has consistently referred to these claims in this particular communication. On the other hand, *Olson* [1969] stated as follows: 'In a series of 13 experiments the molar ratio of retinal formed to β-carotene consumed had an average value of 1.2 with a range of 0.9–1.8. No product other than retinal was detected in this reaction mixture.' It should be noted that the molar ratio of retinal formed to β-carotene consumed was never 2 and that the values varied widely, ranging from 0.9 to 1.8. Later on, *Singh and Cama* [1974] stated as follows: 'The reaction is strictly stoichiometric and 2 mol of retinal are formed upon utilization of 1 mol of β,β-carotene.' But these authors have not given any details of their experimental procedures and of their results.

On reviewing the available information it does appear clear that the entire evidence produced in support of central cleavage has been based on the claims of *Goodman and Huang* [1965] and that the subsequent reports regarding the stoichiometry of the reaction have not been unequivocal. Even the claims of *Goodman and Huang* [1965] were not convincing, because their data were based on the counts of the retinal fraction eluted from the column, while no attempts were made to establish the purity and

Table I. The activities of the 15,15′-dioxygenase isolated from guinea-pig and rabbit intestine

Substrate	Guinea-pig[1]		Rabbit[1]		Rabbit[2]	
	specific activity	relative activity	specific activity	relative activity	specific activity	relative activity
β-Carotene	5.93	1.00	8.41	1.00	0.09	1.00
8′-Apo-β-carotenal	0.56	0.09	0.28	0.03	0.91	10.20
10′-Apo-β-carotenal	3.37	0.57	4.65	0.55	0.96	10.70
12′-Apo-β-carotenal	0.68	0.12	0.57	0.07	–	–

[1] *Singh and Cama* [1974].
[2] *Lakshmanan* et al. [1968].

the actual amounts (mass) of the compound formed. Moreover, it will be seen later on that the rat intestinal mucosa also contains enzymes which can actively oxidize retinal to retinoic acid and can reduce it to retinol, and that both activities are localized in the soluble fraction of the mucosal homogenates. Therefore, we consider it rather unfortunate that such an important point has not received the attention it deserves and that the stoichiometry of the reaction has not been established unequivocally. On the other hand, since retinal has been consistently isolated and identified after incubation with β-carotene, it has been assumed that the attack must have taken place at the central double bond of the carotene molecule. But apocarotenoids also yield retinal in the same system, in which case the formation of retinal from β-carotene via apocarotenoids cannot be ruled out.

In addition to these, there are other points that must be considered and they are as follows: (1) Some of the claims from the different laboratories have been conflicting and these are stated below. (a) *Lakshmanan* et al. [1968] had observed that the rabbit intestine enzyme is about 10 times more active against 8′- and 10′-apo-β-carotenal, as compared to β-carotene, while according to *Singh and Cama* [1974] the enzymes from both guinea-pig and rabbit intestines are markedly more active against β-carotene as compared to 8′-, 10′- and 12′-apo-β-carotenal (table I). (b) While *Goodman* et al. [1967] and *Goodman and Olson* [1969] have stressed the requirement of lipid cofactors, particularly lecithin, *Singh and Cama* [1974] did not mention any requirements of lipids. (2) The central cleavage hypothesis obviously implies extreme specificity for the dioxygenase for the 15,15′ double bond of the carotene molecule and, as has already been discussed, such specificity has not been established unequivocally so far.

V. Central Fission versus Random Fission

Now let us look at the other possible mechanism. It has already been mentioned that after the chemical relationship between β-carotene and vitamin A was established by *Karrer* and his associates it had appeared to them that hydrolytic cleavage at the central double bond of the β-carotene with the addition of two molecules of water should yield two molecules of retinol. On the other hand, on physicochemical grounds *Zechmeister* et al. [1943] had suggested that, because of resonance, the centrally located double bond of a conjugated system should be more stable than the terminal ones, which would imply that terminal attack should be preferred to central fission. Several years later *Pullman and Pullman* [1963] also had, on theoretical considerations, supported the idea that the attack should be terminal and not central.

In contrast to such views, *Hunter and Williams* [1945] and *Hunter* [1946] reported that small amounts of retinal are formed when β-carotene is treated with hydrogen peroxide in glacial acetic acid, while others showed that the yield of retinal can be increased considerably by carrying out the reaction in the presence of a catalyst, osmium tetroxide [*Goss and Macfarlane*, 1947; *Wendler* et al., 1950]. Later on, *Glover and Redfearn* [1954] were able to isolate a whole series of apo-β-carotenals by following the progress of the reaction with time, while more recently by using hydrogen peroxide and potassium permanganate and by controlling the time of the reaction *Hasani and Parrish* [1967] had prepared 8′-, 10′- and 12′-apo-β-carotenal from β-carotene in good yield. It should, however, be mentioned here that, long ago, *Karrer and Solmssen* [1937] had reported that alkaline permanganate attacks the ends of the connecting chain of the conjugated double bonds of β-carotene, giving apo-β-carotenals. It is thus clear that attempts made in several laboratories at effecting chemical fission of the β-carotene molecule by various oxidizing agents led to the formation of more of apo-β-carotenals than retinal.

Early work on the relative growth-promoting activities of α- and γ-carotene had consistently shown that these carotenoids are only half as active as β-carotene, which had supported the idea that the carotene molecule is cleaved centrally during its conversion to vitamin A. But, when pure crystalline vitamin A was similarly compared with β-carotene, it was unexpectedly found to be twice as active on a weight basis. This observation has since been repeatedly confirmed and it is now widely accepted that on a weight basis β-carotene is only 50–60% as active as vitamin A, although isolated claims have been made that the biopotency of 1.0 µg of β-carotene is equivalent to that of 1.0 µg of retinol [*Koehn*, 1948; *Burns* et al., 1951].

If the β-carotene molecule is cleaved centrally on a weight basis, the biopotencies of β-carotene and retinol should be equal. But, since the carotene was found to be actually about half as active as retinol, it follows that half of the carotene molecule is lost during the process of its conversion into vitamin A, and on the basis of such reasonings *Morton* [1940] had suggested that the attack should be asymmetric.

After reviewing the situation at that time *Glover* [1960] had proposed the 'β-oxidation theory', according to which the oxidative cleavage starts at either end of the ethylenic chain with equal probability. The oxidative cleavage then continues with successive removal of two carbon units until the C_{20} unit is reached, when further β-oxidation is blocked by the methyl group located on the C_{13}, which is at the β-position with respect to the central carbon atom of the β-carotene molecule. This is depicted in figure 2.

In his attempts to produce evidence in support of this hypothesis, *Glover* [1960] had isolated from the intestine small amounts of 12′-apo-β-carotenal, 12′-apo-β-carotenol and 12′-apo-β-carotenoic acid ester containing ^{14}C at the 15,15′ positions and 15′-^{14}C-retinol from the liver of vitamin A-deficient rats given a mixture of (15-15′-^{14}C)-β-carotene and unlabelled 12′-apo-β-carotenal.

Another approach to this problem would be to examine the relative biopotencies of different apocarotenoids against β-carotene in curative growth tests with vitamin A-deficient animals, and there has been some work in this direction. Long ago *von Euler* et al. [1938] had prepared 2′- and 4′-apo-β-carotenal by subjecting β-carotene to oxidation with $KMnO_4$ and had found them to be vitamin A-active in rats in dosage levels of 5 and 20 µg, respectively. Later on, *Marusich* et al. [1960] had reported that 8′-apo-β-carotenal was 72% as active as β-carotene in USP curative tests with rats, while *Singh* et al. [1972] had found 6′-, 8′-, 10′- and 12′-apo-β-carotenal to be 38.5, 31.3, 75.9 and 44.2% as active as β-carotene, respectively, in a rat-growth assay. In liver storage tests, where *Hoppe* et al. [1971] had given vitamin A-depleted Japanese quails β-carotene and 8′-apo-β-

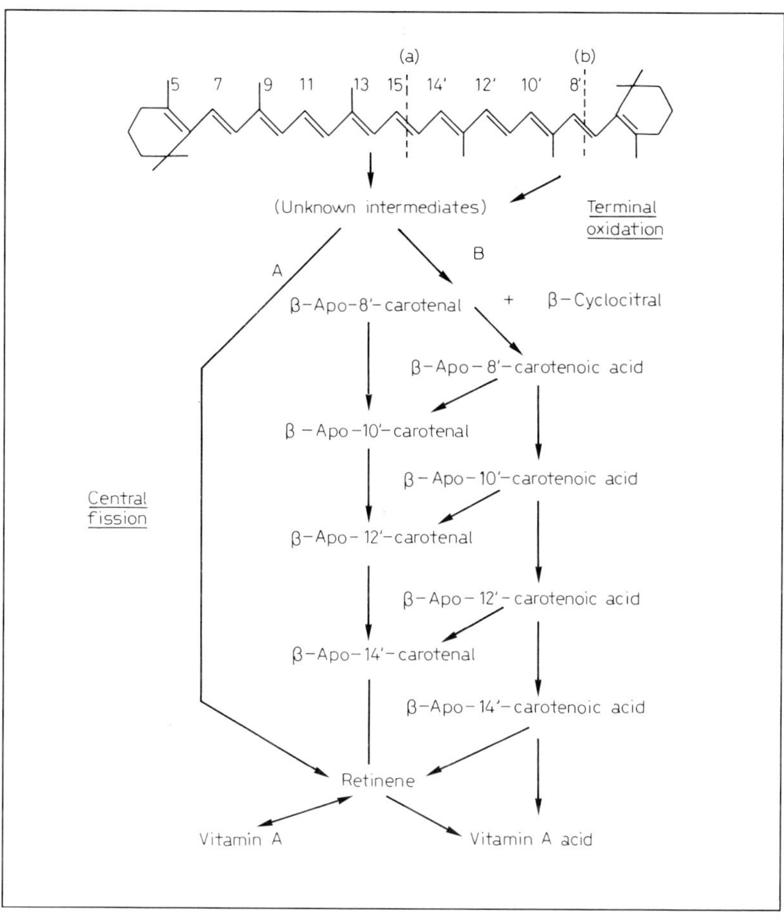

Fig. 2. The pathways of β-carotene metabolism, as suggested by *Glover* [1960].

carotenoic acid ethyl ester in their diet for 14 days, the ethyl ester gave 78 % of the liver vitamin A values given by β-carotene. One major criticism against all this work would obviously be that the biopotencies were compared on a weight basis. The chain lengths of the apocarotenoids vary, but all of them should give one unit of vitamin A. Therefore, it is clear that their activities should be compared on molar basis. At any rate, the relative biopotencies of the apocarotenoids reported by the various groups of workers have not been consistent even on a weight basis and therefore could not lead to any meaningful conclusions.

If the β-carotene molecule and the apocarotenoids were to be cleaved by the dioxygenase reaction at the 15,15′ double bond, then on a molar basis all apocarotenoids should be equally active, while at the same time β-carotene should be twice as active as any apocarotenoid. If, on the other hand, the attack on β-carotene is asymmetric, then on a molar basis it should be as active as the apocarotenoids. On the basis of such reasonings *Sharma* et al. [1976] very carefully evaluated the relative biopotencies of 8′-, 10′- and 12′-apo-β-carotenal against β-carotene as the standard and the respective values found by them were 72, 78 and 72% of the β-carotene value on a molar basis. Since the biopotencies of the apocarotenals were much less than 200% of that of β-carotene, it is clear that the carotene could not have been cleaved centrally.

VI. Metabolism of Retinal, Retinoic Acid and Apo-β-Carotenoids

At this stage it would be pertinent to discuss the metabolism of retinal, retinoic acid and apo-β-carotenoids.

A. Retinal
(1) Liver. Retinal was first discovered in the eye and since then a very large volume of information has been made available regarding its function in the visual system. This particular subject is well known and well-documented and therefore need not be discussed here. But there has been considerable work on the enzymatic formation of retinal from retinol and its further oxidation to retinoic acid. *Zachman and Olson* [1961] had purified 3-fold the alcohol dehydrogenase from the non-particulate supernatant fraction of rat liver homogenates and during the steps of purification the activities against retinol and ethanol were always found in the same fraction. These workers had therefore concluded that the same enzyme is responsible for the reduction of the two aldehydes. *Mahadevan* et al. [1962] had demonstrated that the activity is quantitatively localized in the supernatant fraction of rat liver homogenates, from where it can be precipitated by 45–70% saturation with ammonium sulphate, while very recently *Sharma* et al. [1977] have shown similar quantitative localization of the activity in the soluble fractions of rat liver and chicken liver homogenates.

(2) Intestine. Long ago *Glover* et al. [1948b] had shown that 4–24 h after a dose of retinal, considerable amounts of retinol and retinyl esters,

Table II. Oxidation and reduction of retinal by everted intestinal sacs of rats [adapted from *Deshmukh and Ganguly,* 1967]

	Weight of sac, g	Retinyl-ester, nmol	Retinol nmol	Retinoic acid, nmol	Retinal nmol
Medium	–	–	53.6	113.8	978.5
Sac	1.57	55.3	98.6	120.5	142.7

Table III. Oxidation and reduction of retinal by enzymes of rat intestine [adapted from *Deshmukh and Ganguly,* 1967]

Enzyme source	Cofactor	Retinol nmol	Retinoic acid, nmol	Retinal nmol
Whole homogenate	$+$ NAD$^+$	9.6	24.8	1,496.0
100,000 *g* supernatant fraction	$-$ NAD$^+$	7.0	27.7	1,443.0
	$+$ NAD$^+$	16.1	31.7	1,372.0
	$+$ NADH	14.6	36.0	1,425.0

but not retinal, are found in the intestinal wall of rats. Later on, *Deshmukh* et al. [1965] reported that 6 h after a dose of retinal, significant amounts of retinol and retinyl esters together with large amounts of unchanged retinal are found in the mucosa and muscles of the small intestine of rats. *Deshmukh and Ganguly* [1967] demonstrated reduction of retinal to retinol in the everted intestinal sacs (table II) as well as with the whole homogenates and the particle-free supernatant fraction of the homogenates of rat mucosa (table III). The presence of retinal-reductase activity in rat intestine was soon confirmed by *Fidge and Goodman* [1968], who were able to purify it 13-fold from the supernatant fraction of the mucosal homogenate. The purified enzyme had a molecular weight of 60,000–80,000 and required NADH or NADPH for its activity.

B. Retinoic Acid

(1) Liver. Futterman [1962] had shown that retinal can be oxidized to retinoic acid by the aldehyde dehydrogenase of calf liver. The oxidation and reduction of retinal by rat liver was studied in some detail by *Mahadevan* et al. [1962]. Both activities were quantitatively localized in the supernatant

fraction of the liver homogenate and 0–45% saturation with ammonium sulphate precipitated the oxidase, while 45–70% saturation removed the reductase activity. It was rather surprising that addition of reduced NAD$^+$ markedly stimulated the oxidase activity. More recently *Sharma* et al. [1977] have shown that the oxidase activity is present in the whole homogenates as well as in the particle-free supernatant fraction of the homogenates of rat and chicken liver. Here also, NAD$^+$ and reduced NAD$^+$ markedly stimulated the activities in both species.

(2) Intestine. During their work on the liver enzyme *Mahadevan* et al. [1962] had noticed significant retinal-oxidase activity in the intestine and kidneys of rats. Later on, *Deshmukh and Ganguly* [1967] produced more convincing evidence regarding the presence of the oxidase activity in rat intestine. It has been shown in table II that, following incubation of everted intestinal sacs of rats with retinal in the medium, large amounts of retinoic acid appeared both in the sac and in the medium. Table III has further shown that the oxidase activity is localized in the soluble fraction of the mucosal homogenate and that, here also, it was stimulated by added NAD$^+$ or reduced NAD$^+$. At about the same time *Crain* et al. [1967] reported that the soluble fraction of rat mucosa can oxidize retinal to retinoic acid and that the activity is markedly stimulated by NAD$^+$ or reduced NAD$^+$. Subsequently the mucosal enzyme was purified 160-fold by *Moffa* et al. [1970]. Its molecular weight was calculated to be 80,000 and it contains 2 mol of iron per mol of the enzyme protein.

C. Apo-β-Carotenoids

Apo-β-carotenals have been found to occur extensively in the plant kingdom, though often in very small amounts [*Thommen,* 1971]. In fact, *Thommen and Wiss* [1963] had isolated from commercial alfalfa meals 0.9 mg/kg of 8′-apo-β-carotenal and 4.0 mg/kg of 10′-apo-β-carotenal, while later on *Thommen* [1967] had isolated from fresh alfalfa 6′-, 8′-, 10′- and 12′-apo-β-carotenal as well as retinal. It is widely believed that the apocarotenals are formed through a process of degradation of β-carotene, though the actual mechanism of such degradation of β-carotene has not been demonstrated yet.

Synthetic apocarotenoids are, however, available commercially and are being extensively used in poultry feed for improving the colour of the egg yolk. There has been some work on the metabolism of apocarotenoids, particularly in chickens. Thus, way back in 1951, *Festenstein* had found two carotenoids in horse intestine, which were subsequently identified as 10'- and 12'-apo-β-carotenal [see *Glover,* 1960]. It has already been discussed that *Glover* [1960] was able to isolate small amounts of ^{14}C-labelled 12'-apo-β-carotenal, 12'-apo-β-carotenol and 12'-apo-β-carotenoic acid ester from the intestine of vitamin A-deficient rats given a mixture of ^{14}C-labelled β-carotene and unlabelled 12'-apo-β-carotenal. Similarly *Winterstein and Hegedus* [1960] had detected 8'-apo-β-carotenal in the mucous membrane of rat intestine.

Thommen [1961] had reported that the livers of rats fed 50 mg of 8'-apo-β-carotenal (C_{30}) per day contained significant amounts of the fed apocarotenal and the corresponding apocarotenoic acid. But when rhesus monkeys were given the same apocarotenal, their livers contained, in addition to the fed apocarotenal and the corresponding acid, a C_{24-25} apocarotenoic acid. Later on, *Hasani and Parrish* [1968] had found substantial amounts of vitamin A in the liver, blood serum and egg yolk of Japanese quails given 8'-, 10'- and 12'-apo-β-carotenal, while *Wildfeuer* et al. [1968] had found trace amounts of 8'-apo-β-carotenal, but considerable amounts of the corresponding acid in the egg yolk of hens given 8'-apo-β-carotenal in their feed. In a separate communication *Wildfeuer* [1969] had reported that, when 8'-apo-β-carotenal was included in the poultry feed in the proportion of 10 mg/kg feed, the yolk of the eggs contained two carotenoid metabolites, namely 8'- and 10'-apo-β-carotenoic acid. More recently *Hasani and Parrish* [1972] found considerable amounts of vitamin A in both liver and egg yolk of Japanese quails fed 8'-, 10'- and 12'-apo-β-carotenal. They also identified 10'- and 12'-apocarotenoic acid in the liver and egg yolk of the birds given 10'-apocarotenal, while only the corresponding carotenoic acid and no lower carotenoic acid could be identified in both tissues when 8'-apo-β-carotenal was given.

The important points which have emerged from all these reports can be summarized as follows: (1) Apo-β-carotenals extensively occur in the plant kingdom; (2) these compounds can be formed from β-carotene in the intestine; (3) when given in the diet they are oxidized to the corresponding carotenoic acid, which in turn can be degraded to the lower carotenoic acid, and (4) they give rise to vitamin A when fed to animals.

Against this background the work of *Sharma* et al. [1976, 1977] on apocarotenoids assumes considerable significance. We have already seen that in curative growth tests with vitamin A-deficient rats 8'-, 10'- and 12' apo-β-carotenal were 72–78% as active as β-carotene, on a molar basis [*Sharma* et al., 1976]. The same workers further showed that, when fed to

Table IV. Liver-storage tests of apocarotenoids in vitamin A-deficient rats (the rats were given in 1 week, in three doses, the indicated amounts of the apocarotenoid) [adapted from *Sharma* et al., 1976]

Supplement	Dose/week per rat µmol	Total vitamin A/liver (mean ± SD), nmol
β-Carotene	1	20 ± 3.2
	2	60 ± 5.1
	3	98 ± 6.3
8′-Apo-β-carotenol	2	16 ± 2.9
	4	55 ± 5.5
	6	90 ± 7.2
8′-Apo-β-carotenal	2	8 ± 1.4
	4	25 ± 2.4
	6	40 ± 3.6
8′-Apo-β-carotenoic acid	2	5 ± 0.9
	4	18 ± 1.2
	6	28 ± 3.1

vitamin A-deficient rats, 8′-apo-β-carotenol, 8′-apo-β-carotenal and 8′-apo-β-carotenoic acid ester lead to the deposition of considerable amounts of vitamin A in the liver (table IV). Although the amounts of vitamin A deposited varied markedly, these experiments clearly showed that all of them are cleaved at the 15,15′ double bond.

Sharma et al. [1977] were able to isolate significant amounts of 8′-, 10′- and 12′-apo-β-carotenal together with much larger amounts of retinal from the intestine of chickens given β-carotene (table V). In this work all the apocarotenals and retinal were conclusively identified on the basis of several criteria. It would be interesting to recall here that long ago *Ganguly* et al. [1959] had isolated a pigment from the intestine of chickens given β-carotene. It showed a λ_{max} at 434 nm in light petroleum and its concentrations were 0.9 and 0.72 µg/g of the intestinal mucosae and muscles, respectively. At that time no attempt was made to characterize this pigment and it is possible that it was a similar mixture of several apo-β-carotenals.

Sharma et al. [1977] also isolated appreciable amounts of 8′-, 10′- and 12′-apo-β-carotenoic acid from the intestine of rats given 8′-apo-β-carotenal (table VI). Here also, the three carotenoic acids were conclusively identified on the basis of several criteria.

Table V. 8′-, 10′- and 12′-apo-β-carotenal and retinal isolated from the intestine of White Leghorn male chickens given 6–8 μmol of β-carotene [adapted from *Sharma* et al., 1977]

Metabolite	Absorption maximum in light petroleum, nm	nmol/bird
8′-Apo-β-carotenal	452	0.7
10′-Apo-β-carotenal	435	1.8
12′-Apo-β-carotenal	410	3.6
Retinal	370	7.4

Table VI. Isolation of 8′-, 10′- and 12′-apo-β-carotenoic acid from the intestine of rats given 10 μmol of 8′-apo-β-carotenal [adapted from *Sharma* et al., 1977]

Metabolite	Absorption maxima in light petroleum, nm	nmol/rat
8′-Apo-β-carotenoic acid	444	8.6
10′-Apo-β-carotenoic acid	426	10.2
12′-Apo-β-carotenoic acid	405	13.8

Therefore, the findings of *Sharma* et al. [1976, 1977] were in general agreement with the observations of the previous workers summarized earlier. They further showed that the mitochondrial and microsomal fractions of the homogenates of rat and chicken liver can oxidize apo-β-carotenals to the corresponding carotenoic acids and that addition of NAD^+ or $NADP^+$ leads to a big increase in the oxidase activity in both species.

VII. Possible Mechanism of Conversion

Based on these observations *Sharma* et al. [1977] proposed the mechanism of conversion shown in figure 3.

According to them the dioxygenase non-specifically attacks any one of the double bonds of the β-carotene molecule resulting in the formation of the corresponding apo-β-carotenal or retinal. The apocarotenal formed can in turn be degraded to retinal [*Olson,* 1969] or it can be oxidized to the

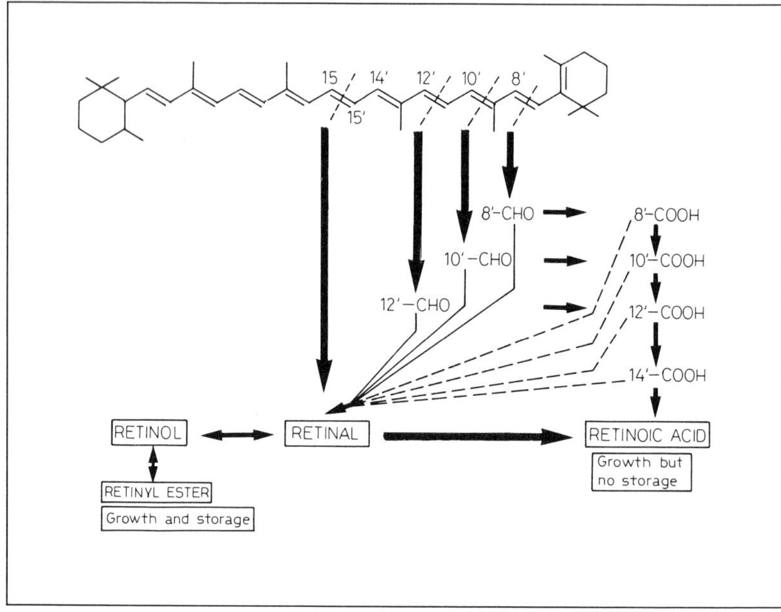

Fig. 3. The mechanism of conversion of β-carotene into vitamin A, as proposed by *Sharma* et al. [1977].

corresponding apocarotenoic acid, which in turn can yield retinal or can be oxidized through a process analogous to β-oxidation ultimately giving retinoic acid. The retinal formed from the various apocarotenoids as well as from β-carotene can be reduced to retinol or it can be oxidized to retinoic acid. The retinol is esterified with fatty acids to retinyl esters which are absorbed and stored in the liver [*Ganguly,* 1960]. The retinoic acid, on the other hand, is not stored in the animal body [*Ganguly,* 1967, 1969], but it can support growth of the animal.

VIII. Concluding Remarks

Since retinal has been isolated and conclusively identified after incubation of β-carotene with enzymes prepared from tissues of several species of animals, it has been generally assumed that the carotene is cleaved at the central double bond in such reactions. But evidence produced regarding the stoichiometry of the reaction has not conclusively established that 1 mol of β-carotene yields 2 mol of retinal. Retinal can be formed from apo-β-carotenals also, in which case the formation of retinal from β-carotene via apocarotenals cannot be ruled out.

Earlier growth tests with vitamin A-deficient rats had shown that β-carotene is only half as active as vitamin A on a weight basis and more recent work has shown that apo-β-carotenals are 72–78% as active as β-carotene, on a molar basis.

The weight of evidence available so far is overwhelmingly in favour of random cleavage and is against any specific attack at the central double bond of the carotene molecule.

References

Bieri, J.G.; Pollard, C.J.: Studies on the site of conversion of β-carotene injected intravenously into rats. Br. J. Nutr. *8:* 32–44 (1954).

Burns, M.J.; Hauge, S.M.; Quackenbush, F.W.: Utilization of vitamin A and carotene by the rat. I. Effects of tocopherol, Tween and dietary fat. Archs Biochem. *30:* 341–346 (1951).

Crain, F.D.; Lotspeich, F.J.; Krause, R.F.: Biosynthesis of retinoic acid by intestinal enzyme of the rat. J. Lipid Res. *8:* 249–254 (1967).

Deshmukh, D.S.; Ganguly, J.: Demonstration of oxidation and reduction of retinal in rat intestine. Indian J. Biochem. *4:* 18–21 (1967).

Deshmukh, D.S.; Murthy, S.K.; Mahadevan, S.; Ganguly, J.: Studies on metabolism of vitamin A: absorption of retinal (vitamin A aldehyde) in rats. Biochem. J. *96:* 377–382 (1965).

Euler, H. von; Karrer, P.; Solmssen, U.: Homologe des Vitamins A (Axerophtols) und ein Abbauprodukt des α-Carotins, α′-Apo-2-carotinal. Helv. chim. Acta *21:* 211–222 (1938).

Fidge, N.H.; Goodman, D.S.: Enzymatic reduction of retinal to retinol. J. biol. Chem. *243:* 4372–4379 (1968).

Fidge, N.H.; Smith, F.R.; Goodman, D.S.: The enzymatic conversion of β-carotene into retinal in hog intestinal mucosa. Biochem. J. *114:* 689–694 (1969).

Futterman, S.: Enzymatic oxidation of vitamin A aldehyde to vitamin A acid. J. biol. Chem. *237:* 677–680 (1962).

Ganguly, J.: Some aspects of the metabolism of carotenoids and vitamin A; PhD thesis, Reading (1949).

Ganguly, J.: Absorption, transport and storage of vitamin A. Vitams Horm. *18:* 387–402 (1960).

Ganguly, J.: Metabolism of vitamin A. J. Sci. industr. Res. *26:* 110–130 (1967).

Ganguly, J.: Absorption of vitamin A. Am. J. clin. Nutr. *22:* 923–933 (1969).

Ganguly, J.; Krishnamurthy, S.; Mahadevan, S.: The transport of carotenoids, vitamin A and cholesterol across the intestine of rats and chickens. Biochem. J. *71:* 756–762 (1959).

Ganguly, J.; Murthy, S.K.: Biogenesis of vitamin A and carotenoids; in Sebrell, Harris, The vitamins; 2nd ed., vol. 1, pp. 125–153 (Academic Press, New York 1967).

Glover, J.: The conversion of β-carotene into vitamin A. Vitams Horm. *18:* 371–386 (1960).

Glover, J.; Goodwin, T.W.; Morton, R.A.: Studies in vitamin A. 8. Conversion of β-carotene into vitamin A in the intestine of the rat. Biochem. J. *43:* 512–518 (1948a).

Glover, J.; Goodwin, T.W.; Morton, R.A.: Studies in vitamin A. 6. Conversion in vivo of vitamin A aldehyde (retinene$_1$) to vitamin A$_1$. Biochem. J. *43:* 109–114 (1948b).

Glover, J.; Redfearn, E.R.: The mechanism of the transformation of β-carotene into vitamin A in vivo. Biochem. J. *58:* XV (1954).

Goodman, D.S.: Biosynthesis of vitamin A from β-carotene. Am. J. clin. Nutr. *22:* 963–965 (1969).

Goodman, D.S.; Huang, H.S.: Biosynthesis of vitamin A with rat intestinal enzymes. Science *149:* 879–880 (1965).

Goodman, D.S.; Huang, H.S.; Kanai, M.; Shiratori, T.: The enzymatic conversion of all-*trans*-β-carotene into retinal. J. biol. Chem. *242:* 3543–3554 (1967).

Goodman, D.S.; Huang, H.S.; Shiratori, T.: Mechanisms of biosynthesis of vitamin A from β-carotene. J. biol. Chem. *241:* 1929–1932 (1966).

Goodman, D.S.; Olson, J.A.: The conversion of all-*trans*-β-carotene into retinal; in Clayton, Methods in enzymology, vol. XV, pp. 462–475 (Academic Press, New York 1969).

Goss, G.C.L.; MacFarlane, W.D.: Oxidation of β-carotene with osmium tetroxide. Science *106:* 375 (1947).

Harashima, K.: Conversion of β-carotene to vitamin A by rat intestinal sections. Biochim. biophys. Acta *90:* 211–213 (1964).

Hasani, S.M.A.; Parrish, D.B.: Procedure for preparing three apo-β-carotenals from β-carotene. J. agric. Fd Chem. *15:* 943–944 (1967).

Hasani, S.M.A.; Parrish, D.B.: Vitamin A activity of β-apo-carotenals in *Coturnix coturnix japonica.* J. Nutr. *94:* 402–406 (1968).

Hasani, S.M.A.; Parrish, D.B.: Forms of vitamin A and carotenoids in tissues, blood serum and yolk of eggs from *Coturnix coturnix japonica* fed β-apo-carotenal. J. Nutr. *102:* 1437–1440 (1972).

Hoppe, P.; Gropp, J.; Tiews, J.: Über die Vitamin A-Wirksamkeit synthetischer Eidotterpigmente bei Hühner- und Wachtelkücken. Arch. Geflügelkd. *35:* 75–76 (1971).

Hunter, R.F.: The conversion of carotene into vitamin A. Nature, Lond. *158:* 257–260 (1946).

Hunter, R.F.; Williams, N.E.: Chemical conversion of β-carotene into vitamin A. J. chem. Soc. *1945:* 554–556.

Karrer, P.; Helfenstein, A.; Wehrli, H.; Wettstein, A.: Pflanzenfarbstoffe. XXV. Über die Konstitution des Lycopins und Carotins. Helv. chim. Acta *13:* 1084–1099 (1930).

Karrer, P.; Morf, R.; Schopp, K.: Zur Kenntnis des Vitamins-A aus Fischtranen II. Helv. chim. Acta *14:* 1431–1436 (1931).

Karrer, P.; Solmssen, U.: β-Carotinal, ein Abbauprodukt des β-Carotins. Helv. chim. Acta *20:* 682–690 (1937).

Koehn, C.J.: Relative biological activity of beta-carotene and vitamin A. Archs Biochem. *17:* 337–344 (1948).

Lakshmanan, M.R.; Pope, J.L.; Olson, R.A.: The specificity of a partially purified carotenoid cleavage enzyme of rabbit intestine. Biochem. biophys. Res. Commun. *33:* 347–352 (1968).

Mahadevan, S.; Murthy, S.K.; Ganguly, J.: Enzymic oxidation of vitamin A aldehyde to vitamin A acid by rat liver. Biochem. J. *85:* 326–331 (1962).

Marusich, W.; DeRitter, E.; Vreeland, J.; Krukar, R.: Vitamin A activity of beta-apo-8′-carotenal. J. agric. Fd Chem. *8:* 390–393 (1960).

Mattson, F.H.; Mehl, J.W.; Deuel, H.J., Jr.: Studies on carotenoid metabolism. 7. The site of conversion of carotene to vitamin A in the rat. Archs Biochem. *15:* 65–73 (1947).

McGillivray, W.A.; Thompson, S.Y.; Worker, N.A.: Further studies on the metabolism by rats of intravenously administered aqueous dispersions of carotenoid pigments. Br. J. Nutr. *10:* 126–134 (1956).

Moffa, D.J.; Lotspeich, F.J.; Krause, R.F.: Preparation and properties of retinal-oxidizing enzyme from rat intestinal mucosa. J. biol. Chem. *245:* 439–447 (1970).

Moore, T.: Vitamin A and carotene VI. The conversion of β-carotene to vitamin A in vivo. Biochem. J. *24:* 692–702 (1930).

Morton, R.A.: The constitution and physiological significance of carotene and allied pigments. Chem. Indust. *59:* 301–307 (1940).

Olcott, H.S.; McCann, D.C.: Carotenase. The transformation of carotene to vitamin A in vitro. J. biol. Chem. *94:* 185–193 (1931).

Olson, J.A.: The conversion of radioactive β-carotene into vitamin A by the rat intestine, in vivo. J. biol. Chem. *236:* 349–356 (1961a).

Olson, J.A.: Absorption of β-carotene and its conversion into vitamin A. Am. J. clin. Nutr. *9:* 1–12 (1961b).

Olson, J.A.: The alpha and the omega of vitamin A metabolism. Am. J. clin. Nutr. *22:* 953–962 (1969).

Olson, J.A.; Hayaishi, O.: The enzymatic cleavage of β-carotene into vitamin A by soluble enzymes of rat liver and intestine. Proc. natn. Acad. Sci. USA *54:* 1364–1370 (1965).

Pullman, B.; Pullman, A.: Quantum biochemistry, p. 435 (Wiley, New York 1963).

Rea, J.L.; Drummond, J.C.: On the formation of vitamin A from carotene in the animal organism. Int. Z. VitamForsch. *1:* 177–183 (1932).

Sexton, E.L.; Mehl, J.W.; Deuel, H.J., Jr.: Studies on carotenoid metabolism. VI. The relative provitamin A activity of carotene when introduced orally and parenterally in the rat. J. Nutr. *31:* 299–319 (1946).

Sharma, R.V.; Mathur, S.N.; Dmitrovskii, A.A.; Das, R.C.; Ganguly, J.: Studies on the metabolism of β-carotene and apo-β-carotenoids in rats and chickens. Biochim. biophys. Acta *486:* 183–194 (1977).

Sharma, R.V.; Mathur, S.N.; Ganguly, J.: Studies on the relative biopotencies and intestinal absorption of different apo-β-carotenoids in rats and chickens. Biochem. J. *158:* 377–383 (1976).

Singh, H.; Cama, H.R.: Enzymatic cleavage of carotenoids. Biochim. biophys. Acta *370:* 49–61 (1974).

Singh, H.; Mallia, A.K.; Cama, H.R.: Metabolism of β-apocarotenals. Biochem. J. *128:* 11p (1972).

Thommen, H.: Metabolism of 8′-apo-β-carotenal. Chimia *15:* 433–434 (1961); cited in Chem. Abstr. *56:* 6520h (1962).

Thommen, H.: Detection and identification of naturally occurring keto carotenoids and carotenals. Int. Z. VitamForsch. *37:* 175–183 (1967).

Thommen, H.: in Isler, Carotenoids, metabolism, pp. 637–668 (Birkhäuser, Basel 1971).

Thommen, H.; Wiss, O.: Isolation of apocarotenals from alfalfa meal. Z. ErnährWiss. *3:* suppl., pp. 18–23 (1963); cited in Chem. Abstr. *60:* 2259h (1964).

Thompson, S.Y.; Braude, R.; Coates, M.E.; Cowie, A.T.; Ganguly, J.; Kon, S.K.: Further studies on the conversion of β-carotene to vitamin A in the intestine. Br. J. Nutr. *4:* 398–421 (1950).

Thompson, S.Y.; Ganguly, J.; Kon, S.K.: The conversion of β-carotene to vitamin A in the intestine. Br. J. Nutr. *3:* 50–78 (1949).

Vartapetian, B.B.; Dmitrovsky, A.A.; Alkhasov, D.G.; Lemberg, I.H.; Girshin, A.B.; Gusinsky, G.M.; Starikova, N.A.; Eropheeva, N.N.; Bogdanova, I.P.: A new approach to the study of the mechanism of biosynthesis of vitamin A from carotene by activation of ^{18}O by the nuclear reaction ^{18}O (α, Nγ)^{21}Ne using cyclotron-accelerated α-particles. Biokhimiya *31:* 881–886 (1966).

Wendler, N.L.; Rosenblum, C.; Tishler, M.: The oxidation of ß-carotene. J. Am. chem. Soc. *72:* 234–239 (1950).

Wiese, C.E.; Mehl, J.W.; Deuel, H.J., Jr.: Studies on carotenoid metabolism. 8. The in vitro conversion of carotene to vitamin A in the intestine of the rat. Archs Biochem. *15:* 75–79 (1947).

Wildfeuer, I.: Metabolites in egg yolk after feeding hens 8′-apo-β-carotenal. Z. Lebensmitt. UntersForsch. *140:* 140–144 (1969); cited in Chem. Abstr. *71:* 99289b (1969).

Wildfeuer, I.; Acker, L.; Mehner, A.; Rauch, W.: Effect of dietary carotenoid supplementation on egg yolk colour. III. Addition of 8′-apo-β-carotenal, ethyl 8′-apo-β-carotenoate, canthaxanthin and paprika. Z. Lebensmitt. UntersForsch. *136:* 203–214 (1968); cited in Chem. Abstr. *68:* 103065h (1968).

Winterstein, A.; Hegedus, B.: Detection of biologically active aldehydes. Chimia *14:* 18–19 (1960); cited in Chem. Abstr. *54:* 21455f (1960).

Woolf, B.; Moore, T.: Carotene and vitamin A. Lancet *223:* 13–15 (1932).

Zachman, R.D.; Olson, J.A.: A comparison of retinene reductase and alcohol dehydrogenase of rat liver. J. biol. Chem. *236:* 2309–2313 (1961).

Zachman, R.D.; Olson, J.A.: The uptake of ^{14}C-β-carotene and its conversion into retinol ester (vitamin A ester) by the isolated perfused rat liver. J. biol. Chem. *238:* 541–546 (1963).

Zechmeister, L.; Le Rosen, A.L.; Schroeder, W.A.; Polgár, A.; Pauling, L.: Special characteristics and configuration of some stereoisomeric carotenoids including prolycopene and pro-γ-carotene. J. Am. Chem. Soc. *65:* 1940–1951 (1943).

J. Ganguly, PhD, DSc, FNA, Department of Biochemistry,
Indian Institute of Science, Bangalore 560012 (India)

Subject Index